A PERFECT THREAT

POPULATION GROWTH, CLIMATE CHANGE & NATURAL RESOURCE DEPLETION IN THE 21st CENTURY

CRAIG R. SMITH, M.D.

Ordering Information:
Quantity sales. Special discounts are available on quantity purchases by corporations, associations, and others. Orders by US trade bookstores and wholesalers. For details, contact the publisher at the address above.

Editing by The Pro Book Editor
Interior and Cover Design by IAPS.rocks

ISBN (paperback): 978-1-7368094-1-9
ISBN (hardcover): 978-1-7368094-2-6
        Main category—Science/Global Warming & Climate Change
        Other category—Nature/Natural Resources

First Edition

This work is dedicated to my grandsons, William and Thomas Olson.

CONTENTS

PREFACE

W E LIVE IN EXTRAORDINARY TIMES. The world has changed in ways that few could have imagined just 75 years ago. In one lifetime alone the population of the world increased from about 3 billion to almost 8 billion people and world gross domestic product increased from about $9 trillion to $85 trillion.[1,2] To support this rapid population and economic growth, world energy consumption increased by 500%, from about 100 exajoules (EJ) in 1950 to nearly 600 EJ in 2020.[3] Burning coal, oil, and natural gas provided most of this energy and increased global fossil fuel-derived carbon dioxide emissions from about 5 billion tons per year to nearly 40 billion tons.[4] These emissions produced a fundamental change in the Earth's fast carbon cycle and caused an increase in the concentration of carbon dioxide in the atmosphere from about 290 parts per million (ppm) in 1900 to more than 410 ppm in 2020, levels that are unprecedented in the last 800,000 years.[5,6] As predicted, this increase in atmospheric carbon dioxide was sufficient to alter the Earth's energy balance and produce an increase in the global mean surface temperature of about 1°C (1.8°F) relative to the 20th century

1   *World Population Prospects*. United Nations, Department of Economic and Social Affairs, Population Division (2019), https://population.un.org.

2   The World Bank, World Development Indicators (2020). GDP (constant 2010 US$), http://data. worldbank.org.

3   British Petroleum Company. *BP Statistical Review of World Energy 2020* (London: British Petroleum Co., 2020), http://BP.com.

4   Hannah Ritchie and Max Roser, *CO_2 and Greenhouse Gas Emissions* (2017), https://ourworldindata. org.

5   NOAA Global Monitoring Laboratory. Carbon Cycle Greenhouse Gases (2020), https://Esrl.noaa. gov.

6   Parts per million (ppm) refers to the concentration of a substance. It is a dimensionless number that is generally used to describe a low concentration of a substance with a big effect. One ppm is equal to 1 mg/kg (1 milligram per kilogram) by weight or 1 mL/L (1 milliliter per liter) by volume. To put this number in perspective, 1 ppm is about 1 grain of sugar in 275 sugar cubes or 7 drops of ink in a bathtub of water.

average, with further warming expected during the remainder of this century.[7,8] Global warming caused major environmental changes, including melting of polar and glacial ice[9], changes in global cloud cover[10], a sea level rise of more than 6 inches[11], and alterations in the temperature and acidity of the Earth's oceans.[12] Depletion of natural resources such as fossil fuels, fresh water, coral reefs, and arable land occurred coincident with these changes, driven by the same basic forces, population and economic growth.

None of these facts are new. For decades demographers, climatologists, and geologists have studied population growth, climate change, and natural resource depletion and urged international cooperation aimed at mitigating the serious adverse consequences of these developments. For example, in 2009 the Population Institute published a brief article entitled "2030: The Perfect Storm Scenario."[13] This article related population growth to climate change and natural resource depletion. The following is an excerpt:

"The year is 2030 and the "perfect storm" forecast by John Beddington, England's chief scientific advisor, has arrived. These are the key developments:

- Consistent with earlier demographic projections, global population has increased from 6.8 billion to 8.3 billion.
- Demand for food has grown by 40 percent, but supplies have not kept up with demand. World grain reserves are empty. The FAO (the Food and Agriculture Organization of the United Nations) warns that bad harvests and grain export embargoes are pushing 200 million people to "the brink of starvation."
- Demand for energy has grown by 45 percent, but supplies in recent years have not kept up, and the resulting scarcity has pushed energy prices to record highs.

---

7   GISTEMP Team, *GISS Surface Temperature Analysis (GISTEMP), version 4*, NASA Goddard Institute for Space Studies (2020). Dataset accessed 2020-10-24 at https://data.giss.nasa.gov/gistemp/.

8   N. Lenssen, G. Schmidt, J. Hansen, M. Menne, A. Persin, R. Ruedy, and D. Zyss, "Improvements in the GISTEMP uncertainty model," *J. Geophys. Res. Atmos.*, 124, no. 12 (2019): 6307-6326, doi:10.1029/2018JD029522.

9   N. Wonderling, M. Willeit, J. F Donges, and R. Winklemann, "Global warming due to the loss of large ice masses and Arctic summer sea ice," *Nature Communications* 11 (2020): 5177, https://doi.org/10.1038/s41467-020-18934-3.

10   J. R. Norris, R. J. Allen, A. T. Evan, M. D. Zelinka, C. W. O'Dell, and S. A. Klein, "Evidence for climate change in the satellite cloud record," *Nature* 536 (2016): 72-75, https://doi.org/10.1038/nature18273.

11   T. Slater, A. E. Hogg, and R. Mottram, "Ice-sheet losses track high-end sea-level rise projections," *Nat. Clim. Chang.* 10 (2020): 879–881, https://doi.org/10.1038/s41558-020-0893-y.

12   Andrew J. Pershing et al., "Challenges to natural and human communities from surprising ocean temperatures," Proc. Natl. Acad. Sci., (2019): DOI, 10.1073/pnas.1901084116.

13   https://www.foresightfordevelopment.org/sobipro/55/933-2030-the-perfect-storm-scenario.

- Global demand for water has grown by close to 30 percent. Water conservation measures have helped, but close to four billion people are now living in an area of high water stress.
- Efforts to reduce greenhouse gas emissions have fallen far short of the pledges made in 2010. Climate change is more evident, and the concentration of greenhouse gas emissions is dangerously close to the 450-ppm level that could trigger the worst effects of climate change.
- Shortages of food, energy and water are increasing the number of failed states and ratcheting up international tensions."

When I first read this article I recalled "peak oil"[14] and *The Population Bomb*[15] but thought these issues were behind us. Oil prices had fallen dramatically, natural gas was abundant, and most developed countries had decreasing population growth rates. Climate change was widely discussed in the media, but I understood the Paris Climate Accords and new technology would likely limit global warming to a "manageable" 2°C or less this century.

What had I missed? If John Beddington and the Population Institute were correct, why weren't policy makers and the media discussing the impact of population growth or natural resource depletion in addition to climate change? Why weren't thought leaders considering how population growth, climate change, and natural resource depletion interact and planning strategies to mitigate this interaction?

My initial curiosity turned into a strong interest. I needed to know more. I needed to understand the science and the data underlying these projections. I began spending the majority of my time studying the science and evaluating the evidence from many different fields, including demography, environmental science, climatology, and geology. I found a wealth of information publicly available, including primary data obtained by academic scientists, private institutions, and governmental agencies. The peer-reviewed literature was abundant and wide-ranging. Innumerable articles, editorials, blogs, and commentary were available.

My strong interest became a profound concern. It became apparent that the relationship between population growth, climate change, and natural resource depletion was extremely important but received little attention outside of academic and public policy circles. The threat wasn't immediate, but it was extraordinary: by the end of the 21st century, continued population and economic growth will cause depletion of many essential natural resources

---

14   "Peak oil" refers to the work of M. King Hubbert, who in 1956 postulated that the world's oil supplies would go into irreversible decline. This work led to concerns that rapid depletion of fixed petroleum reserves would lead to soaring costs, strict rationing, and conflict.

15   *The Population Bomb* was a 1968 best seller written by Paul and Anne Ehrlich that warned of the consequences of overpopulation, including famine, environmental degradation, and social unrest.

and produce serious deterioration in many terrestrial and marine habitats, including our own.[16]

Unfortunately, most people have not begun to understand the nature of this threat: how these forces interact, how they vary geographically, and why mitigating them requires an unprecedented and coordinated reallocation of capital on a global scale. Some continue to see the issues as a "debate" or even a political issue. The truth is that the cumulative long-term effect of these forces is not a debate. It is a fact. It is serious, and it is difficult to accept. Ignorance, intellectual arrogance, politics, nationalism, or special interests simply can't be allowed to obfuscate the threat. We can't let our "tribal" instincts prevent us from taking the steps necessary to forge a sustainable future for our grandchildren.

As my understanding of population growth, climate change, and natural resource depletion grew, I decided that I should share what I learned with the general public and write this book. I am not a demographer, climate scientist, geologist, or engineer. I am a physician and entrepreneur trained in medicine, statistics, and epidemiology. I have spent the better part of my life evaluating evidence and translating it into various plans or policies. Over many years, I developed the ability to translate complex data and concepts into a form that can be understood by most people. These are the skills and the perspective I have brought to the task of writing this book. Some of the data and concepts are complex. There are no apologies for this level of complexity. The book requires thought and patience, but it does not require technical or scientific expertise.

The book will take you on a guided tour of scientific data, identify associated uncertainties, and describe the inferences that can be reasonably drawn. You will explore the range of mitigation strategies that are currently available or are under development. At the end of this journey, you will have a much more in-depth understanding of population growth, climate change, natural resource depletion, and the strategies that can be employed to mitigate their adverse consequences.

Footnotes have been used liberally throughout the text. Most of these footnotes refer to primary sources, but some provide a definition or additional information designed to clarify the text. Here are a few additional suggestions for reading this book:

1. If you find an unfamiliar technical term, first check the Glossary. It will likely contain a helpful definition. If not, use your internet search engine to look for a definition of the term or concept you need help with.

---

16 C. Xu, T. A. Kohler, T. M. Lenton, J.-C. Svenning, M. Scheffer, , "Future of the Human Climate Niche," *Proc. Natl. Acad. Sci.*, 117(21) (2020): 11350-11355, www.pnas.org/cgi/doi/10.1073/pnas.1910114117.

2. If you find an unfamiliar unit of measurement, check the Units of Measurement at the back of the book. A description of the metric should be found there.

3. There is a brief summary at the end of each chapter that is designed to provide the "take away" messages for readers and can be used as a quick reference.

4. I can be contacted at APerfectThreat@gmail.com. I will do my best to answer any questions you may have and would be interested in your comments about the book.

It is my expectation that anyone reading this book will evaluate the evidence presented and challenge the conclusions in an effort to formulate their own thinking about these threats and mitigation strategies. Once equipped with this understanding, it is my hope readers will take action in their personal lives and hold their governments and political leaders accountable for pursuing the bold task of changing course.

Several data resources have been used liberally as the basis for this work. Prominent among these sources are scientific journals, such as *Science, Nature*, and other peer-reviewed journals. *The World Bank* website and the website for *Our World in Data* were used frequently. The website for the UN Department of Economic and Social Affairs, Population Division; the CIA Factbook; the Pew Research Center; and the website PopulationPyramid.com were the source of much of the data on population growth. The websites for NASA, especially for the Goddard Institute for Space Studies (GISS); the National Oceanic and Atmospheric Administration (NOAA); and the Massachusetts Institute of Technology (MIT) Joint Program on the Science and Policy of Climate Change were major sources of data and information for the chapter on climate change. The chapter on natural resource depletion relied on the *June 2020 BP Statistical Review of World Energy*, Exxon Mobil's *2020 Energy and Carbon Summary,* and the websites for the U.S. Energy Information Administration (EIA) and International Energy Agency (IEA). These resources are used many times in the book. Additional resources are cited for individual graphs and tables or for key scientific articles published in peer-reviewed journals.

This work is entirely my own. I am responsible for any errors or misstatements. I have not received any direct or indirect compensation, grants, or gifts for writing the book. I hope you will judge it to be objective and apolitical. I would like to acknowledge the assistance and encouragement of many friends and colleagues. There were times I put this work aside, and I needed their support and encouragement to finish. Finally, I would like to express my deep appreciation to my wife, Susan, for her tolerance and understanding throughout the long and tortuous process of assembling this volume.

Craig R. Smith, M.D.

CHAPTER 1

Storms of the Century

"Hope is not a strategy."

EVIDENCE HAS BEEN ACCUMULATING FOR the past several decades that mankind may be threatened by its own existence sometime in the next 100 years.[17] The clouds on the horizon are darker than ever and portend a disturbing forecast: humanity and the world we live in are headed directly into three immense and enormously powerful storms in the latter half of the 21st century. These storms are human population growth, global climate change, and progressive natural resource depletion. These storms have been gradually growing in intensity, converging, and interacting with each other. Collectively they are having a profound impact on the environment and the world economy. Some geographic regions have already been severely affected, while some others have been relatively spared. How mankind responds and adapts to these storms will to a large extent determine the course of history.

The broad sweep of history has been shaped by remarkably few primary factors. Among these factors are the availability of natural resources, climatic conditions, and our human proclivity for the formation of social networks or tribes. Tribalism is a deeply ingrained human behavior that likely provided a survival benefit to our distant ancestors.[18] In many ways, economic, social, and political history has been determined by tribal efforts to control access to natural resources, the degree to which various tribes have been able to adapt to the environment, our pursuit of intertribal or intratribal social or economic dominance, and mankind's ability to develop and use technology to achieve these ends. Over the millennia these needs, proclivities, and abilities have to a large extent determined the course of history. Time and time again the details and circumstances have changed, but the fundamental dynamics have remained similar. If this premise is true, our

---

17 S. C. Sherwood and M. Huber, "An adaptability limit to climate change due to heat stress," *Proc. Natl. Acad. Sci.* 107(21) (2010): 9552-9555, https://doi.org/10.1073/pnas.0913352107.

18 C. J. Clark, B. S. Liu, B. M. Winegard, and P. H. Ditto, "Tribalism Is Human Nature," *Curr. Dir. Psych. Sci.* 28(6) (2019): 587-592, https://doi.org/10.1177/0963721419862289.

collective response to changes in the availability of natural resources and/or changes in the environment, including our use of existing tools or development of new technology, will largely determine our fate.

Over the past several decades each of these impending storms has been studied and characterized by the international scientific, energy, environmental, and/or public health communities.[19,20] The path each of these storms will likely take has been forecast and widely discussed in scientific meetings, public hearings, publications, and the popular media.[21,22] The convergence and interaction of these storms has been explored and widely discussed. As these storms grow in intensity, the sense of urgency felt by governments, policy makers, and the public waxes and wanes, depending on the near-term forecast, media coverage, and the relative urgency of other issues. Certain influential individuals or entities that see the forecast as disadvantageous to their interests work to minimize and obfuscate the evidence and discredit proposed abatement strategies.

The public is generally aware that global temperatures are rising, and that the climate is slowly changing, but many are either ignorant of the threat, choose to disregard the risk, or have been misled into believing that these changes are due to nothing more than "normal" climatic cycles. Awareness of the impact of population growth and natural resource depletion is much less prevalent, and there is nearly a complete lack of public appreciation of the consequences of the interaction of these "storms" or their geographic distribution. In the last 20 years these issues have been superseded by concern over carbon emissions, global warming, and emerging infectious diseases. The most recent Gallup survey found that American's views about global warming have changed little since 2001; only 64% of Americans believe global warming is caused by human activity; and only 45% believe global warming will pose a serious threat in their lifetime.[23]

On an international scale, public awareness and acceptance of the threat posed by anthropogenic carbon emissions is growing, especially in industrialized countries. Unfortunately, public opinion regarding the causes and severity of the risk has been inconsistent. A 2018 Pew Research Center survey found that more than 80% of the

---

19   J. Reilly et al., *Food• Water• Energy• Climate Outlook, Perspectives from 2018* (2019), http://global-change.mit.edu.

20   MIT Joint Program on the Science and Policy of Global Change, *2018 Food Water Energy and Climate Outlook* (2018), https://globalchange.mit.edu.

21   J. Hansen, *Storms of My Grandchildren: The Truth About the Coming Climate Catastrophe and Our Last Chance to Save Humanity* (New York, NY: Bloomsbury, 2009).

22   T. Palmer, and B. Stevens, "The Scientific Challenge of Understanding and Estimating Climate Change," *Proc. Natl. Acad. Sci.*, 116.49 (2019): 24390-24395.

23   L Saad, "Americans as Concerned as Ever About Global Warming" (2019), https://News.gallup.com.

population of Greece, South Korea, France, Spain, and Mexico believe climate change is a major threat, but less than 60% of those living in the US, South Africa, Indonesia, Russia, and Nigeria believe the threat from climate change is serious. In these latter countries 15-20% of the population indicated that climate change is not a threat at all and 20-25% see it as a minor threat.[24]

A similar Pew Research Center survey conducted in the US in 2016 found that more than 30% of respondents think climate change is due to "natural causes" and less than 50% believe it is due to "human activity."[25] More than 50% believe that "technology" will solve most problems associated with climate change. In the US, the public and political leaders have become so galvanized in their views about climate change that meaningful dialogue has become nearly impossible. Among Republicans with a medium level of science knowledge only 25% report believing human activity causes climate change while 71% of Democrats with the same level of science knowledge believe it does.[26] Some people have incorrectly interpreted any level of uncertainty about the methods or data underlying the climate forecast to mean that the forecast is a hoax, perpetrated by unscrupulous scientists and environmentalists.

Humans respond best to threats that are personal, sudden, immoral, or nearby.[27] We respond most strongly to news that is shocking or novel, even when it's false. We deal poorly with future threats, especially when they develop gradually. We also tend to ignore objective data, especially if we see the data as irrelevant or self-evident, or we don't understand it. In general, we see the future the way we would like it to be and typically pick evidence that supports our preconceived expectations. We also have the propensity to believe bad things will happen to other people and have enduring optimism about our own immunity from future threats. We handle uncertainty poorly and often subscribe to the wisdom of sources that reflect our personal social or political norms. We tend to agree with the social group (or "tribe") we align ourselves with. Social media have taken advantage of this tribal vulnerability and greatly amplified it.

Miscalculation is another historical theme that reflects mankind's propensity to ignore the obvious or to distort data if it doesn't serve our interest. Poor judgment, acting precipitously without all the facts, distorting facts to serve a preconceived purpose, and

24   M. Fagan and C. Huang, "A look at how people around the world view climate change," (2019), https://www.pewresearch.org.

25   Fagan and Huang, "Climate change." https://www.pewresearch.org

26   M. Brenan and L. Saad, "Global Warming Concern Steady Despite Some Partisan Shifts" (2019), https://news.gallup.com.

27   G. Marshall, *Don't Even Think About It: Why Our Brains Are Wired to Ignore Climate Change.* (New York, NY: Bloomsbury, 2014).

following inept or misguided leadership are actions or circumstances that have, from time to time, gotten mankind in a lot of trouble. These behavioral traits make the threats facing humanity in the 21st century particularly problematic. They are collectively a perfect threat because they have profound consequences, yet humanity is ill equipped to recognize the threat or respond effectively to it.

Humans are inherently curious and seek to gain an understanding of the world in which they live through exploration, observation, and experimentation. These methods provide evidence that can be weighed, evaluated, and used to answer questions and create knowledge that promotes understanding.

Navigation through the pitfalls and uncertainty often encountered in the pursuit of the truth is best achieved through critical thinking.[28] Critical thinking is based on reason and logic but is not an inherent human skill; it is learned.[29] It is a tool through which reasoned conclusions can be drawn. Critical thinking is objective and rational; it is not emotional or political. It is an iterative process that is open to revision based on new data or new analysis. Critical thinking is an approach that can be used to decide whether a claim is likely to be true, partially true, or false. It is to be distinguished from flawed or biased forms of thinking, such as anecdote, dogma, speculation, prejudice, propaganda, and deception. Data and analyses that are published in peer-reviewed journals are especially valuable because they have been subject to scrutiny by experts. Data or conclusions that have not been subject to quality and peer review should be given less weight. Use of critical thinking is absolutely essential in order to understand population growth, climate change, and natural resource depletion, their interaction, their geographic distribution, and their consequences.

As you will see, the time has passed when we might have been able to chart a more desirable course in order to completely avert these impending storms. It is difficult to see how technology or even humanity's remarkable adaptive capacity will allow us to completely escape serious consequences.[30] An immediate change in human reproductive behavior and an unprecedented, coordinated global reallocation of resources is necessary to alter our current course.

There is hope that world leaders will quickly develop and implement effective measures that will permit mankind to survive the height of the convergent storms and allow us

---

28   "Our Concept and Definition of Critical Thinking" (2019), https://criticalthinking.org.

29   "Critical Thinking" (2018), https://www.plato.stanford.edu.

30   M. Bologna and G. Aquino, "Deforestation and World Population Sustainability: A Quantitative Analysis," *Sci Rep* 10 (2020): 7631, https://doi.org/10.1038/s41598-020-63657-6.

to emerge into a sustainable world in the 22nd century. There is also hope that major advances with existing technology or new "breakthrough" technologies will mitigate these threats and that serious political, social, and economic consequences will not materialize. Unfortunately, hope is not a strategy.

CHAPTER 2

Human Population Growth

"Population growth is the primary cause of environmental damage."
—Jacques Cousteau

Chapter Guide

THIS CHAPTER DESCRIBES HUMAN POPULATION growth in some detail and explains the basics of population dynamics. The sources of population data and the level of uncertainty surrounding these data are discussed. The reader should understand the many ways future population growth can be forecast and especially the underlying assumptions of the most frequently used projection, the UN Population Division Medium Variant Projection. At the end of the Chapter you should be familiar with a population pyramid and understand the dynamics of population growth, including where it is occurring and why. You should understand some of the main consequences of population growth, including urbanization, transnational migration, and environmental degradation. Finally, the chapter discusses the relationship between population growth and economic growth. A summary concludes the chapter.

Introduction

Current evidence suggests modern humans (*Homo sapiens*) evolved from earlier species of the genus *Homo* around 300,000 years ago in Africa.[31] Among the many advantageous anatomical and functional changes that evolved over time are a bipedal erect posture, a large brain, and language. These and other evolutionary traits enabled modern humans to adapt to exogenous threats, to occupy a very broad environmental niche, and to compete with other species for natural resources. There were several other species of the genus *Homo* that coexisted with early humans, including Neanderthals in Europe and Denisovans in Asia. Multiple lines of evidence, including genetic analyses, seem to confirm that early humans migrated out of Africa at least 88,000 years ago to other regions of the world,

---

31   E. Callaway, "Oldest *Homo sapiens* Fossil Claim Rewrites Our Species' History," *Nature* (2017), https://doi.org/10.1038/nature.2017.22114.

possibly in multiple waves, crossbreeding to some extent along the way with other closely related *Homo* species, including Neanderthals and Denisovans.[32]

One of the more noteworthy of our species' evolutionary changes is the ability of females to reproduce throughout the year rather than only during estrus cycles. Human females are also able to become pregnant shortly after giving birth but do not exhibit any overt signs of fertility that can be readily detected by males. Human males and females have different body characteristics and vary in size and appearance. These and other traits form the biological basis for human sexual behavior and the considerable reproductive capacity of humans. A consequence of this reproductive capacity has been a dramatic increase in the human population over the last 300,000 years (Figure 2.1).[33]

Figure 2.1: Estimated world population from 10,000 BCE to present.

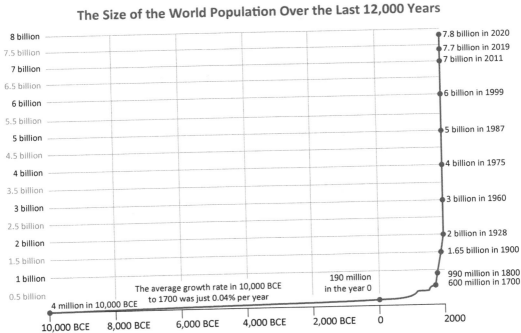

The Size of the World Population Over the Last 12,000 Years

For the majority of the past 300,000 years the rate of human population growth was low, probably less than 0.05% per year, and at times may have actually been negative. Development of new agricultural technology about 10,000 years ago resulted in major societal changes that facilitated incremental increases in the rate of human population growth. As a consequence,

32   T. Rito, D. Vieira, M. Silva, et al., "A Dispersal of *Homo sapiens* from Southern to Eastern Africa Immediately Preceded the Out-of-Africa Migration," *Sci Rep* 9 (2019): 4728, https://doi.org/10.1038/s41598-019-41176-3.

33   M. Roser, H. Ritchie, and E. Ortiz-Ospina, "World Population Growth" (Published online at OurWorldInData.org, 2013), https://ourworldindata.org/world-population-growth.

the world human population at the dawn of the 19th century C.E. is estimated to have been 900 million ± 150 million people.[34] With the advent of the industrial revolution, human population growth greatly accelerated and has been exponential over the past 200 years. This exponential growth is a relatively recent development and has been driven largely by cheap energy and technological advancement, especially in agricultural technology.

Population growth is a function of the rate of new live births per woman of childbearing potential, the number of women of childbearing potential, and the rate people are dying. Exponential human population growth is due to a rapidly falling death rate and a more gradual reduction in the birth rate. (Figure 2.2)

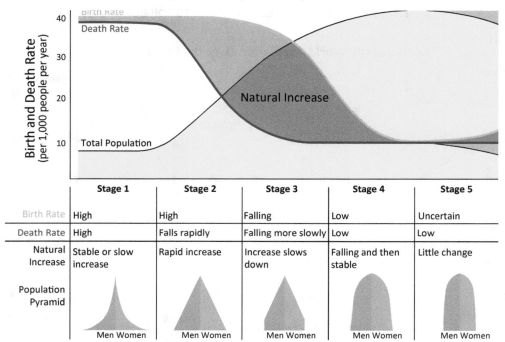

Figure 2.2: Population Dynamics

	Stage 1	Stage 2	Stage 3	Stage 4	Stage 5
Birth Rate	High	High	Falling	Low	Uncertain
Death Rate	High	Falls rapidly	Falling more slowly	Low	Low
Natural Increase	Stable or slow increase	Rapid increase	Increase slows down	Falling and then stable	Little change
Population Pyramid					

Even if total fertility rates decline, population growth can continue because of population momentum. The number of women of childbearing potential determines population momentum. A population pyramid depicts the percent of the population by gender in each 5-year increment from birth to 100+ years. If the number of women who reach childbearing potential (age >15 years) in a year exceeds the numbers who no longer have childbearing potential (age >49 years), the absolute number of women of childbearing potential increases. Even if the total fertility rate remains unchanged, there will still be population growth because there will be more women of childbearing potential. This is positive population

34   J. Durand, *The Modern Expansion of World Population.* Proceedings of the American Philosophical Society, *111*(3) (1967): 136-159, http://www.jstor.org/stable/985711.

momentum. Of course, population momentum can be negative if the number of women reaching childbearing potential is fewer than the number who can no longer have children.

Data Sources and Methods

Demography is the study of human populations. Demographers use a variety of methods to measure the size, structure, and distribution of human populations and any spatial or temporal changes that may occur. Demographers use direct and indirect methods to measure and characterize populations. Direct methods include vital statistic registries maintained by governments (or other local or regional institutions) and periodic census taking in which governments attempt to enumerate all of the people living in their country during one brief period in time.

Census data may include collection of data on age, gender, place of residence, educational level, occupation, religion, citizenship, place of origin, and language spoken. After a census is taken, there is usually an analysis done to estimate the accuracy of the census. Most countries conduct a census every 8-10 years. A few collect data more frequently and some countries in the developing world have not conducted a census in many years, especially if there has been civil strife. Methods vary and may be unreliable. Indirect census methods involve various sampling techniques, such as surveys, that can be used to estimate the characteristics of various population groups or the general population. Administrative databases, such as health records, also contribute to understanding changes in population demographics.

Population projections are based on data gathered by diverse methods in more than 190 countries around the world and can be biased by political or other country-specific considerations. In some countries, extra-governmental agencies conduct periodic surveys from which general population data can be inferred. In 1998 the US National Academy of Sciences conducted an extensive review of the methods used to gather population data worldwide and concluded that world population projections until the year 2050 are based on sound scientific evidence and provide plausible forecasts of demographic trends for the world. Short-term world population projections are generally considered to have an error rate of less than 5%. The panel cautioned, however, that projections for specific countries, for certain population groups, or for longer periods are less certain.

From these direct and indirect data sets, demographers estimate a variety of rates and ratios that characterize the changes they observe. Examples of these rates are birth rates, fertility rates, death rates, replacement rates, life expectancy, and many others. In each case, there is some level of uncertainty about the estimates for the expected values. Demographers also use mathematical models to characterize populations and changes that occur over time. Examples of these models include life tables, proportional hazard models, population projections, and population pyramids or population momentum models.

The United Nations Population Division publishes *World Population Prospects*.[35] This annual projection for world population includes assumptions based on past experience. The basic assumptions include country-specific mortality rates, fertility rates, migration patterns, and population momentum. The UN defines the "medium variant" projection as follows: "Medium-variant projection: in projecting future levels of fertility and mortality, probabilistic methods were used to reflect the uncertainty of the projections based on the historical variability of changes in each variable. The method takes into account the past experience of each country, while also reflecting uncertainty about future changes based on the past experience of other countries under similar conditions. The medium-variant projection corresponds to the median of several thousand distinct trajectories of each demographic component derived using the probabilistic model of the variability in changes over time. Prediction intervals reflect the spread in the distribution of outcomes across the projected trajectories and thus provide an assessment of the uncertainty inherent in the medium-variant projection." There are many other "variants" reported by the UN, as described in Table 2.1.

Table 2.1: Projection variants and the assumptions
for fertility, mortality, and international migration

Projection variants	Assumptions		
	Fertility	Mortality	International Migration
Low fertility	Low	Normal	Normal
Medium (fertility)	Medium (based on median probabilistic fertility)	Normal (based on median probabilistic fertility)	Normal
High fertility	High	Normal	Normal
Constant-fertility	Constant as of 2015-2020	Normal	Normal
Instant-replacement-fertility	Instant-replacement as of 2020-2025	Normal	Normal
Momentum	Instant-replacement as of 2020-2025	Constant as of 2015-2020	Zero as of 2020-2025
Constant-mortality	Medium	Constant as of 2015-2020	Normal
No change	Constant as of 2015-2020	Constant as of 2015-2020	Normal
Zero-migration	Medium	Normal	Zero as of 2020-2025

Current Data

In mid-2020 the world population was estimated to be approximately 7.8 ± 0.4 billion people and is projected by the UN Population Division using its medium fertility variant projection to grow to 9.7 ± 0.6 billion by the year 2050, an increase of 2.0 billion people in the next 30 years.[36] The world population growth rate in 2020 is estimated to be 1.03%. The estimated world population since 1950 and the corresponding rate of world population growth for the medium fertility variant are shown in Figure 2.3. For the past 40 years the world population growth rate has declined and is projected to continue to decline through the end of the century. By 2050 the population growth rate is projected to decrease to approximately 0.5% and by 2100 it is projected to decline even further to 0.1%.[37] However, despite the projected decline in the world population growth rate, the world population in 2100 is forecast to reach 11.0 ± 1.6 billion, driven largely by population momentum in the developing world. If the decline in the population growth rate is more gradual than predicted, the world population could exceed 12 billion in 2100. Conversely, if the growth rate declines more rapidly, the world population could be 10 billion or a little less.

Figure 2.3: World population growth from 1950 to 2100; medium fertility variant

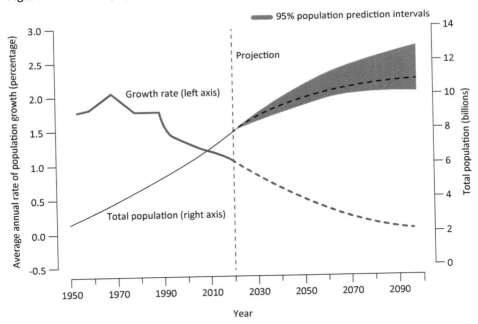

In 2019 the world birth rate was greater than the death rate and there were more women of childbearing potential, resulting in an increase in the world population of about 82.4 million people. However, for the last 20 years both the death rate and the birth rate have

36   United Nations, Department of Economic and Social Affairs, Population Division. *World Population Prospects 2019* (2019).

37   D. E. Bloom, "7 Billion and Counting," *Science* 333 (2011): 562-569.

been in decline, as shown in Figure 2.4. During this period, the decline in the annual death rate was 0.69% per year, while the decline in the birth rate was 0.94% per year. Because the rate of decline in the birth rate exceeded the rate of decline in the death rate, the rate of total population growth has decreased.

Figure 2.4: World birth rate (yellow) and death rate (red) from 1960 to 2017

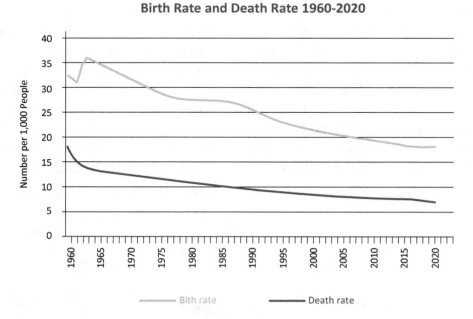

Birth Rate and Death Rate 1960-2020

Reduction in the world birth rate is believed to be because of urbanization, economic growth, improved education, increased availability of contraceptives and abortion, and the empowerment of women to make reproductive choices in some societies. Factors that can sustain high birth rates in the developing world are poverty and economic dependence on children for support in later life, certain cultural beliefs, lack of education, and lack of reproductive healthcare for women.[38] Cultural beliefs include entrenched religious tenets or traditional practices. The reduction in the world death rate is believed to be due to improved sanitation, nutrition, and health care, resulting in less malnutrition, effective treatments for many infectious diseases, and reduced childhood mortality. Population replacement rates are currently about 2.33 live births per woman of childbearing age for the world and 2.1 live births per woman of childbearing age in the developed world. In the developing world, the replacement rate is estimated to vary between 2.5 to 3.3 live births per woman of childbearing age, depending on differences in national or regional mortality rates for females of childbearing potential. Figure 2.5 shows the population pyramid for the world in 2019.[39]

---

38   R. Lee, "The Outlook for Population Growth," *Science* 333 (2011): 569-573.

39   https://populationpyramid.net.worldpopulation.

Figure 2.5: The population pyramid for the world

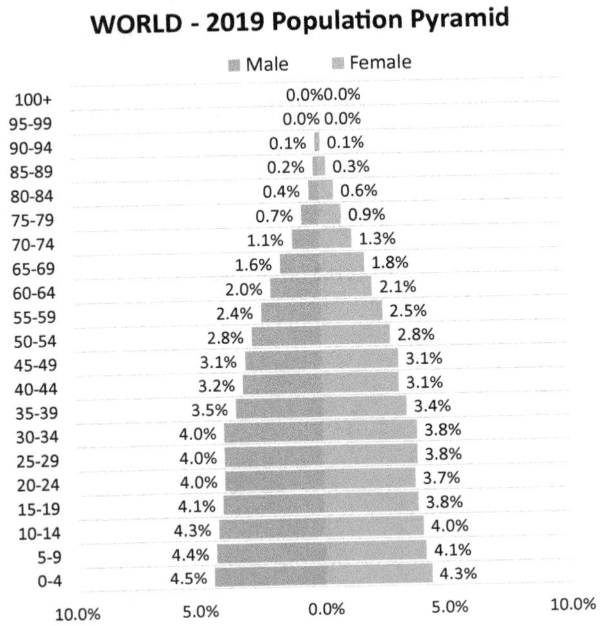

WORLD - 2019 Population Pyramid

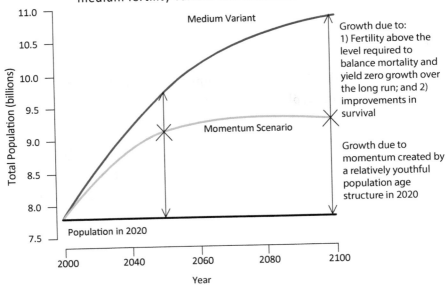

As shown, the percent of females age 10-14 (4.0%) exceeds the percent age 45-49 (3.1%), so the world currently has positive population momentum. The influence of population momentum is large and very important. If global fertility rates were to fall to replacement levels immediately and mortality rate estimates remain unchanged, world population would still increase, as shown in Figure 2.6.

Figure 2.6: Projected size of the world population 2020 to 2100, medium fertility variant and momentum scenarios

Population dynamics vary greatly from country to country, including total fertility rates, mortality rates, and population momentum. The sum total of these effects is a net country-specific population growth rate that determines future country-by-country demographics. Figure 2.7 is a map of the world with 2015-2020 country-specific total fertility rates depicted.[40] Rates are number of live births per 1000 women, assuming all women live to the end of their childbearing years.

Figure 2.7: Total fertility rates by country, 2015-2020

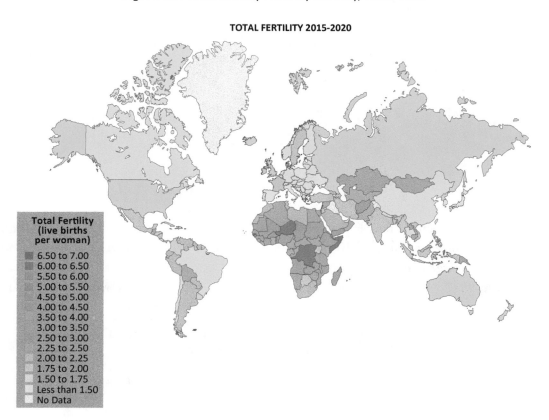

Africa, Southern Asia, and Latin America account for most of the countries in which the total fertility rate exceeds the replacement level. As a result, as shown in Figure 2.8, the expected growth in world population will occur unevenly.[41]

Almost all of the projected growth in world population though 2050 will occur in Africa, India, and Asia (Table 2.2).[42] Population growth will occur primarily in lower latitudes (25°N to 25°S), in the Eastern hemisphere, and will occur in regions where the predominant religion is Islam.

---

40   https://population.un.org/wpp/Maps/.

41   https://population.un.org/wpp/Maps/.

42   "Urban Population (percent of total population)" (2019), https://data.worldbank.org.

Figure 2.8: Average annual population growth rate by country

Rate of
change (%)

4 to 5
3 to 4
2 to 3
1 to 2
0 to 1
Less than -1
No Data

Table 2.2: Mid-year 2019, 2050, and 2100 global population estimates

Region	2019 Population Estimate (Billion)	2019 Population Growth Rate (Percent/ Year)	2050 Population Estimate (Billion)	2100 Population Estimate (Billion)	2050 Estimated Population Density (People/mi^2)
Asia	4.60	0.9	5.29	4.72	439
Africa	1.31	2.5	2.49	4.28	221
Europe	0.75	0.1	0.71	0.63	84
Latin America	0.65	0.9	0.76	0.68	100
North America	0.37	0.7	0.43	0.49	60
Oceania	0.04	1.4	0.06	0.07	47
World	7.71	1.1	9.74	10.87	195

Within Asia population growth will occur primarily in India and the Middle East. (Table 2.3)[43] The UN includes Iran in South Asia, while Georgia, Azerbaijan, and Armenia are included in Western Asia.

---

43   UN Department of Economic and Social Affairs, Population Division, *World Population Prospects, 2019.*

Table 2.3: Within Asia regional population growth estimates from 2019 to 2100

Region of Asia	Main Countries	2019 Population Estimate (Billion)	Current Population Growth Rate (Percent/Year)	2050 Population Estimate (Billion)	2100 Population Estimate (Billion)
Western	Middle East	0.275	1.6	0.383	0.419
Eastern	China/Japan	1.673	0.4	1.617	1.223
Southern	India/Pakistan	1.918	1.2	2.40	2.219
South Eastern	Indonesia	0.662	1.1	0.794	0.744
Central	Kazakhstan	0.073	0.6	0.100	0.115
Total for Asia		4.60	0.9	5.29	4.72

North America, Russia, China, and Europe will experience little or no population growth beyond 2050, other than through immigration. Russia, China, and Western Europe will actually see population declines after 2050 because of negative population momentum and declining total fertility rates. These projections assume that governments in these regions do not provide incentives for women to have more children in order to promote economic development.

Figure 2.9 shows the 2018 population pyramid for China and Europe. As can be seen, in the next few years there are fewer females entering their childbearing years in China (2.5%) than there are females who will no longer be able to bear children (4.0%). A similar picture is seen in Europe.[44] Furthermore, the total fertility rate in China is 1.7, well below replacement, and is currently 1.6 in Europe. The resultant shift in age distribution will influence the potential for economic growth.

Religious beliefs are a source of cultural differences. Christianity is the largest religious group in the world with about 2.4 billion followers, representing 31% of the world's population. Islam is second with about 1.9 billion adherents or 24% of the world's population. Agnostics and Hindus follow with about 800 million each. This distribution is changing because the population growth rate among Muslims is 1.5% per year compared with a 0.7% population growth rate for all other religions and agnostics combined.[45] The high population growth rate among Muslims is because of a high total fertility rate, estimated to be about 2.9 live births per woman of childbearing potential, and positive population momentum in Muslim countries. By comparison, the total fertility rate in non-Muslim countries is 2.1 live births per woman of childbearing potential.

---

44   https://populationpyramid.net.worldpopulation.

45   M. Lipka and C. Hackett, "Why Muslims Are the World's Fastest Growing Religious Group" (2017), https://pewresearch.org.

Figure 2.9: 2018 population pyramids for China and Europe

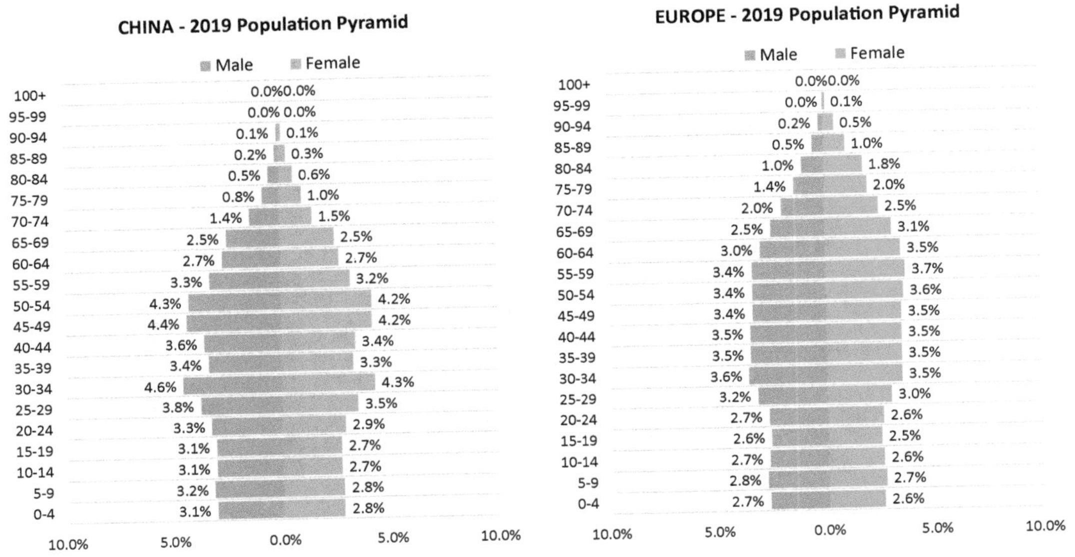

CHINA - 2019 Population Pyramid

■ Male ■ Female

	Male	Female
100+	0.0%	0.0%
95-99	0.0%	0.0%
90-94	0.1%	0.1%
85-89	0.2%	0.3%
80-84	0.5%	0.6%
75-79	0.8%	1.0%
70-74	1.4%	1.5%
65-69	2.5%	2.5%
60-64	2.7%	2.7%
55-59	3.3%	3.2%
50-54	4.3%	4.2%
45-49	4.4%	4.2%
40-44	3.6%	3.4%
35-39	3.4%	3.3%
30-34	4.6%	4.3%
25-29	3.8%	3.5%
20-24	3.3%	2.9%
15-19	3.1%	2.7%
10-14	3.1%	2.7%
5-9	3.2%	2.8%
0-4	3.1%	2.8%

10.0% 5.0% 0.0% 5.0% 10.0%

EUROPE - 2019 Population Pyramid

■ Male ■ Female

	Male	Female
100+	0.0%	0.0%
95-99	0.0%	0.1%
90-94	0.2%	0.5%
85-89	0.5%	1.0%
80-84	1.0%	1.8%
75-79	1.4%	2.0%
70-74	2.0%	2.5%
65-69	2.5%	3.1%
60-64	3.0%	3.5%
55-59	3.4%	3.7%
50-54	3.4%	3.6%
45-49	3.4%	3.5%
40-44	3.5%	3.5%
35-39	3.5%	3.5%
30-34	3.6%	3.5%
25-29	3.2%	3.0%
20-24	2.7%	2.6%
15-19	2.6%	2.5%
10-14	2.7%	2.6%
5-9	2.8%	2.7%
0-4	2.7%	2.6%

10.0% 5.0% 0.0% 5.0% 10.0%

There are several direct consequences of these global demographic changes, including urbanization, transnational migration, and environmental degradation.

Urbanization

Individuals and families in developing countries often migrate from rural areas to regional urban centers in search of economic opportunity and the essentials of life. Once in an urban setting, fertility rates for migrants tend to drop to levels consistent with the urban setting in which they now reside, presumably because children are an economic burden in urban settings and women have greater access to education and economic opportunity in cities. Because of rural-to-urban migration, the world urban population has been growing at a much more rapid pace than the rural population since 1960 (Figure 2.11).[46]

Figure 2.10 depicts the population pyramids for Africa, India, Pakistan, Iraq, Indonesia, and Egypt.[47] The population pyramids demonstrate that the aforementioned countries (and the continent of Africa) have very large cohorts of young women who will enter their childbearing years over the next decade. Combined with a fertility rate that is much greater than replacement, these demographic characteristics indicate that population growth in these countries will continue well into the 21st century, even if fertility rates decline during this period.

---

46 Hannah Ritchie and Max Roser, "Urbanization," published online at OurWorldInData.org (2019), https://ourworldindata.org/urbanization.

47 https://populationpyramid.net.worldpopulation.

Figure 2.10: Population pyramids for Africa, India, Pakistan, Iraq, Egypt, and Indonesia

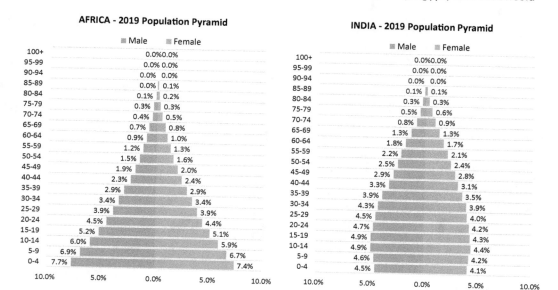

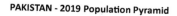

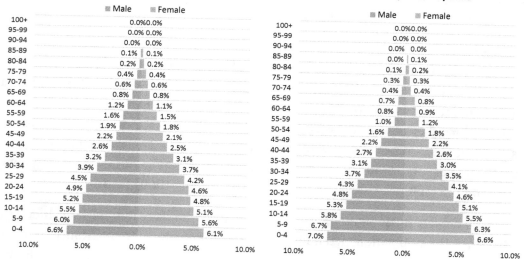

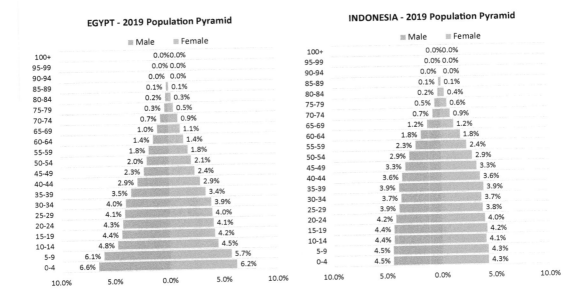

EGYPT - 2019 Population Pyramid

INDONESIA - 2019 Population Pyramid

In 2020 about 56% of the world's population lived in cities.[48] Because of rural to urban migration, it is estimated that by 2050 about 70% of the world's population will live in cities with more than one million inhabitants. By 2100, more than 20 cities will likely have a population greater than 50 million, and a few cities could have more than 75 million inhabitants. Cities projected to have more than 50 million inhabitants by 2100 include Lagos (Nigeria), Kinshasa (Zaire), Dar-Es-Salaam (Tanzania), Nairobi (Kenya), Mogadishu (Somalia), Mumbai (India), Kolkata (India), Delhi (India), Kabul (Afghanistan), Lahore (Pakistan), and Manila (Philippines). Almost all of these "megacities" will be located in Africa or India, where dense population will likely place a heavy burden on infrastructure and local government. If the need for housing, water, and public health is not met, large areas of substandard housing without essential utilities such as electricity, sewage, and water management (i.e., slums) could become even more common. Many countries in the developing world will likely have more than 30% of their population living in slums (Figure 2.12).

Access to basic healthcare, including reproductive healthcare, will continue to be very limited. Unemployment or underemployment could exceed 50% unless means to provide economic opportunity are developed. These cities could become breeding grounds for disease, crime, and opportunists, and enclaves in some of the largest cities in the developing world could be ungovernable. Lawful economic opportunity will likely be unavailable or nonexistent. The value of human capital will decrease. Desperation could

---

48   UN Department of Economic and Social Affairs, Population Division, "World Urbanization Prospects, 2018," https://population.un.org.

force some inhabitants to seek a better life through any available means. Alternatively, national and local governments will have the resources and infrastructure to provide necessary services, and mechanisms will be developed to provide economic opportunity and basic healthcare, including birth control. If governments take these actions, then the adverse consequences of urbanization can be mitigated, at least to some extent.

Figure 2.11: Change in the rural and urban population of the world from 1960 to present

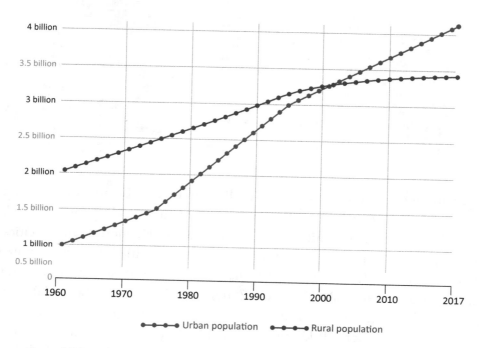

Transnational Migration

There are currently about 270 million transnational migrants in the world (Figure 2.13).[49] A transnational migrant is someone who decides to move from one country to another in search of a better life. A refugee is someone who must leave a country because of an imminent threat. Refugees are protected by international agreements. Migrants are not, although the 2018 UN Marrakesh Compact provides nonbinding international standards for the treatment of transnational migrants.

---

49   United Nations, Department of Economic and Social Affairs, "Trends in International Migrant Stock: Migrants by Destination and Origin," (United Nations database, POP/DB/MIG/Stock/Rev.2017), http://www.un.org/en/development/desa/population/migration/data/estimates2/estimates17.shtml.

Figure 2.12: Share of urban population living in slums

URBAN POPULATION LIVING IN SLUMS 2014

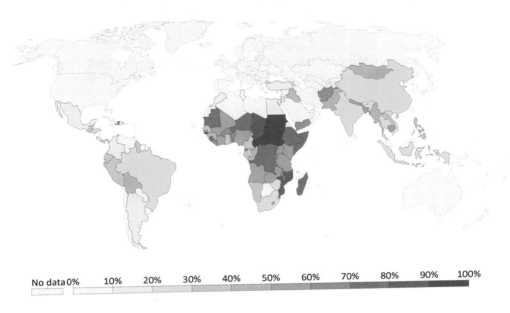

No data 0%   10%   20%   30%   40%   50%   60%   70%   80%   90%   100%

It seems likely that current migratory patterns will persist as long as the countries to which migrants emigrate do not erect effective barriers.[50] Regions or countries likely to experience heavy immigration pressure include North America, Europe, and Australia. Europe is especially vulnerable because there are no major natural barriers to migrants from Africa and the Middle East. The Atlantic and Pacific Oceans are barriers to immigration into North America, although these barriers can be overcome by air or sea travel. The border of the United States and Mexico is the only major land route available to migrants into Canada and the United States. Oceans and seas are barriers to migration into Australia. Several countries have already erected physical and legal barriers to immigration and can be expected to continue to do so if immigration is considered a threat to domestic tranquility, political stability, and local culture. For example, India has built fencing along its border with Bangladesh, the US has built fencing along its southern border with Mexico, and Australia has adopted strict policies to prevent illegal immigration.

Canada, Northern Europe, and Russia may be particularly attractive to immigrants if the climate warms more than 2°C to 3°C in the northern hemisphere. Regions likely to generate large numbers of migrants are sub-Saharan Africa, the Middle East, India, and

---

50   Rainer Münz, *Demography and Migration: An Outlook for the 21st Century* (Washington, DC: Migration Policy Institute, 2013).

Indonesia. If migration is exacerbated by more extreme climate change, and regions such as the Mediterranean basin, equatorial South America, Central America, and Southeast Asia also become major sources of migrants, the social and political pressure on North America and Northern Europe could reach a tipping point, resulting in widespread political unrest.

Figure 2.13: Estimated total international migrant stock from 1960 to 2019

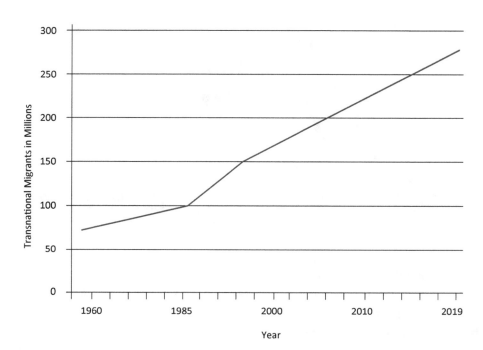

There has already been a dramatic increase in the number of Muslims living in Europe.[51] The Pew Research Center estimates that the Muslim population of Europe was 25.8 million in 2016 and growing rapidly because of continued immigration, an influx of Syrian refugees, and a much higher birth rate among resident Muslims compared to the native European population. In 2016 Muslims represented about 5% of Europeans, but the Pew Research Center estimates this proportion could grow to 14% or more by 2050.[52] More recent data suggest that 8-9% of the population of France, Sweden, Austria, Belgium, and the Netherlands are Muslim compared with less than 1.5% of the population in 1970. Muslim immigrants tend to cluster in cities. There are several European cities in which Muslims account for 15% or more of the population and within these cities there are enclaves where more than 35% of the population is Muslim.

51   C. Hackett, "5 Facts about the Muslim Population in Europe" (2017), http://pewrsr.ch/2i3Tlim.

52   Pew Research Center, ,"Europe's Growing Muslim Population" (2017), https://www.pewforum.org.

The majority of Muslim immigrants assimilate well.[53] However, it is difficult for some Muslims to adapt to European customs because some Islamic sects expect adherents to live their daily lives according to dictates that are contrary to many European laws or cultural values. This cultural clash can lead to alienation, isolation, and resentment among migrant Muslims, and animosity, bigotry, and discrimination on the part of Europeans. This clash has already led to violence and political unrest.

Africa has about 2.5 million immigrants each year, most of whom are from Europe or Asia. In contrast, about 18 million migrants leave the continent each year, most going to Europe (~6.5 million) or Asia (~5.5 million). However, most immigration occurs within Africa. Almost 20 million Africans migrate between African countries each year.[54]

Asia, including India, has over 60% of the world's population but less than 10% of the world's immigrants. Japan is a largely homogenous society as a result of strong cultural, geographic, and language barriers. Immigrants, very few of whom are non-Asian, represent less than 1.5% of Japan's population. India is similar with immigrants from Bangladesh, Pakistan, and Nepal accounting for about 1% of the population. Southeast Asian workers tend to migrate between neighboring countries, especially where there are ineffective barriers. There are more than 1 million Indonesian workers in Malaysia and more than 1 million Myanmar workers in Thailand. In these cases, workers move from a poorer or more repressive country to a richer or more liberal neighboring country.

China is much more culturally diverse and has a significant indigenous Muslim population that has been living in the northwest of the country for centuries. Within China there are significant legal and cultural barriers ("hukou") to regional migration. Despite these limitations, there is still significant rural-urban migration within China as the populace seeks greater economic opportunity. More than 20% of Chinese live away from their place of registration, limiting their access to some governmental benefits and services.

In 2017 China had about 1 million immigrants, representing less than 0.07% of the population. Most of these immigrants are Chinese from Hong Kong or Macau, or Koreans. In 2016 China granted permanent residence to only 1,500 migrants. In contrast, the US has more than 37 million legal immigrants, representing more than 10% of the population. Illegal immigration into the US adds another 12 million or so, bringing the

53   R. Inglehart and P. Norris, "Muslim Integration into Western Cultures: Between Origins and Destinations," HKS Faculty Research Working Paper Series RWP09-007, John F. Kennedy School of Government, Harvard University (2009), http://nrs.harvard.edu/urn-3:HUL.InstRepos:4481625.

54   World Economic Forum, "African Migration: What the Numbers Really Tell Us," (2018), https://www.weforum.org/agenda/2018/06/heres-the-truth-about-african-migration.

total number of immigrants to about 15% of the total US population. The immigrant population of Russia is about 8%, Germany 15%, Canada 22%, and Australia 28%.

Many Asian countries send workers abroad. China and India received a third of the $400 billion in cash transfers from developed countries to developing countries in 2012. Chinese workers migrate to South Korea, Japan, Europe, and North America as well as to African and Latin American countries. Indian information technology workers represent an important outflow of professionals, but most Indian, Bangladeshi, and Pakistani migrants are low-skilled workers employed in countries such as Saudi Arabia and the United Arab Emirates. The Philippines sends more workers abroad than any other Asian country, almost 1.5 million a year, including 300,000 who are employed on the world's ships. About 10 percent of Filipinos live abroad, including 40 percent who are temporary foreign workers. Cash transfers to the Philippines from abroad are equivalent to 10 percent of Philippine GDP.

Currently, immigrants represent about 3% of the world's population. If climate change and natural resource depletion drive this percentage up to as little as 6% and the world population is >10 billion before the end of the century, there could be more than 600 million migrants worldwide.[55,56] Add the impact of prolonged regional conflicts and political repression, and the numbers quickly reach staggering levels by today's standards. Then add the effect of climate change and natural resource depletion, which could make some regions of Earth nearly uninhabitable. Modern civilization has never had to adapt to such massive population movements, and it is far from clear what the political, economic, and social consequences of such relocation could be.

Environmental Degradation

Population growth is associated with depletion of natural resources, loss of arable land, deforestation, pollution, and loss of biodiversity. Population growth is one of a number of human-related factors that can affect the environment. Other human-related (i.e., anthropogenic) factors include civil conflict, wars, polluting technologies, or distortionary governmental policies. The degree to which population growth causes environmental damage or the degree to which it contributes with other factors to adverse environmental effects is subject to debate. The relationship is complex and involves many factors other than population growth, such as use of technology, human behavior, and economic factors, such as trade and political agreements.

---

55 R. Black, W. N. Adger, N. W. Arnell, S. Dercon, A. Geddes, and D. Thomas, "The Effect of Environmental Change on Human Migration," *Global Environmental Change*, Vol. 21, Suppl. 1 (2011): S3-S11, https://doi.org/10.1016/j.gloenvcha.2011.10.001.

56 Philip Martin, "The Global Challenge of Managing Migration," *Population Bulletin* 68, no. 2 (2013), www.prb.org.

Some sociologists and economists have asserted that human ingenuity (through the increased supply of more creative people) and market substitution (as certain resources become scarce) will avert future resource crises due to overpopulation. In this line of thinking, market failures and inappropriate technologies are more responsible for environmental degradation than population size or growth, and technology can be used to substitute for depleted natural resources. Whether or not population growth or human behavior is responsible for environmental degradation is an important policy issue. Dissecting the role of human behavior from the impact of population size can be a difficult and complex undertaking. No doubt there is a strong and direct association between population growth and environmental degradation. No doubt human behavior and the choices humans make, including their use of technology, contribute to this effect. Thus, controlling population growth and changing human behavior will be necessary to lessen environmental degradation. Addressing one and not the other of these effects will likely be a less effective approach.

Economic Growth

The relationship between population growth and economic growth is complex. The size of an economy, or Gross Domestic Product (GDP), is usually measured by the total value of goods and services produced during a specified time period. Value is measured in some commonly accepted international currency (e.g., the US dollar), and comparisons are made using some reference value of the currency (e.g., 2010 US dollars) to adjust for inflation.

There are several methods used to calculate GDP. The most commonly used method sums personal expenditure, government spending, business investment, and the balance of trade. Thus, the size of an economy is determined by the size of the population and the level of personal spending, the amount of government spending, the amount of private capital invested, and the balance of trade with other countries. The level of productivity of the labor force is dependent on the level of technological development within the economy, the degree of specialization within the labor force, the business sectors driving growth, and the availability of natural resources.

Dividing GDP by the population size and calculating GDP *per capita* removes the effect of population growth and can be used as a measure of the aggregate effect of the other factors driving growth. Growth in *per capita* GDP could be due to an increase in spending per person, or it could be due to an increase in government spending, an increase in business investment, or a favorable shift in the balance of payments on a *per capita* basis. World *per capita* GDP has increased over the last 70 years (Figure 2.14). The US has enjoyed steady growth for this entire period, as have European countries and the other OECD countries. China and India did not begin to experience significant growth until the 1980s, but the rate of growth in these economies for the last 25-30 years has outpaced growth in the US and other OECD economies. Africa, on the other hand, has experienced little growth in *per capita* GDP since 1970.

Figure 2.14: Per capita GDP in 2011 US$ from 1950 to 2017

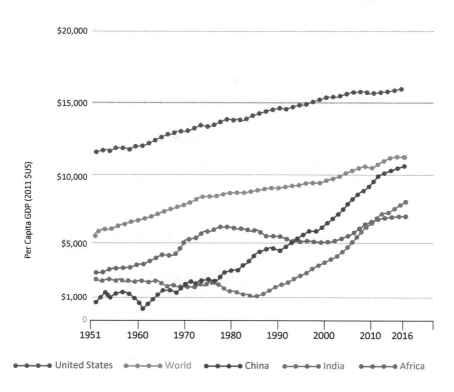

In high-income, urban economies, population growth is associated with growth in per capita GDP, likely indicating specialization in the labor force, increased capital investment, and possibly indicating an increase in the standard of living. In low-income, agricultural economies, population growth is associated with a reduction in *per capita* GDP because unemployment increases or the productivity of the labor force does not increase with population growth. In general, high population growth rates in high-income countries increase *per capita* economic growth while high population growth rates in low-income countries lower *per capita* economic growth, especially if the labor force is unskilled, agriculture dominates the economy, educational opportunities are limited, capital for investment is scarce, and innovative technology is not available.

Population growth rates in OECD countries, China, and Russia are projected to be low or even negative over the next 50 years or longer. Low population growth in these countries would be expected to lower economic growth, although other factors could offset this effect. Population growth in Africa and Indonesia would not be expected to increase economic growth and could cause a decrease in *per capita* GDP in the coming decades. The outcome in India and the Middle East will depend on the degree to which the growing population enters the labor force and labor productivity increases. Furthermore, these relationships would also suggest that migration from low-income countries to high-income countries would increase world economic growth if and only if migrants enter the labor force and are productive.

Summary

The number of humans on Earth has increased exponentially over the last 200 years. Exponential population growth is driven by cheap energy and technological advancement, and is a result of rapidly falling death rates and a more gradual reduction in birth rates. The current world population is estimated to be 7.8 ± 0.4 billion people and is expected to grow to 9.7 ± 0.6 billion by the year 2050. By 2100 world population will likely be 10-12 billion. The world currently has positive population momentum, which will continue for at least the next 15-20 years. Even if total fertility rates drop to replacement levels, the total world population will continue to increase because of population momentum.

World population growth will occur primarily in Africa, the Middle East, and India. Population growth will occur primarily in lower latitudes (25°N to 25°S), in the Eastern hemisphere, and will occur in regions where the predominant religion is Islam. North America, Russia, China, and Europe will experience little or no population growth beyond 2050, other than through immigration. Russia, China, and Western Europe will likely see population declines after 2050 because of negative population momentum and declining total fertility rates. By 2050 it is likely that more than 70% of the world's population will live in cities with more than 1 million inhabitants. By 2100 African Muslims will represent about 40% of all working age people on Earth, and there will likely be more than 600 million transnational migrants worldwide. Europe, North America, and Australia are the most likely destinations for many of these migrants.

Because humans use large amounts of energy and natural resources for producing food and economic growth, population growth will likely continue to cause environmental degradation. Urbanization and transnational migration could lead to increased cultural conflicts and social unrest. Economic growth in developed countries will likely slow as a result of low population growth. Per capita GDP in Africa is likely to decrease as a result of population growth. Many countries are now or will be stressed by population growth and/or transnational migration. The effectiveness of government, the rule of law, regulatory quality, and corruption control could deteriorate, especially in countries that do not have a history of stable political institutions.

CHAPTER 3

Climate Change

"Men argue. Nature acts."
—Voltaire

Chapter Guide

THIS CHAPTER IS AN INTRODUCTION to the science and environmental impact of climate change. An appreciation of the Earth's energy budget and carbon cycle is important since perturbation of these two fundamental natural cycles is the cause of climate change. The sources of climate data and the level of uncertainty surrounding these data are discussed in the chapter, including the uncertainty regarding climate sensitivity. The reader should understand the buffering effect of the oceans on future global warming. The potential magnitude and geographic distribution of future climate changes is discussed in detail. At the end of the Chapter you should understand the dynamics of climate change, including where it is occurring and why, and the main environmental consequences of climate change, including global warming. A summary concludes the chapter.

Introduction

Over the millennia, humans have demonstrated a remarkable ability to adapt to a wide range of climatic conditions. Development of specialized clothing and shelters, harnessing various forms of energy, and the development of agriculture are examples of the technology humans have used to make this adaptation possible. Cultural adaptations have also played a role, such as the formation of cooperative social groups and our ability to pass on information to subsequent generations. These capabilities have allowed humans to live and thrive in a wide range of conditions, from arctic deserts to equatorial jungles. Some scientists have suggested that humans are among the most adaptive species on Earth, largely because of our ability to find creative solutions to problems, employ tools, and cooperate socially. Climate and weather have also played a central role in human history. Civilizations have collapsed based on climatic changes such as prolonged droughts, severe storms, floods, and "little" ice ages. Wars have been won and lost based on the weather.

Modern humans have greatly benefitted from our adaptive capacity and from our ability to innovate, but humanity is still vulnerable to climatic changes and severe weather.

It is important to distinguish between climate and weather. "Climate" is the long-term pattern of temperature, precipitation, humidity, barometric pressure, cloud cover, and wind. The time period used to evaluate climatic conditions is usually 30 years or more. "Weather," on the other hand, describes the short-term pattern of these same conditions. Generally, the climate of a terrestrial region is related to its latitude, altitude, terrain, land cover, and proximity to oceans and their currents.

"Climate change" refers to variation in global or regional climate. Changes in the weather do not necessarily indicate that a change has occurred in a region's climate. A cold winter or a warm summer does not, as an isolated occurrence, necessarily indicate that any change has occurred in climate. Only when these changes persist over periods of 10 or 20 years or more do they indicate a climate change may have occurred. "Global warming" refers to an increase in global mean surface temperatures over 30 years or more and is a climate change. Other climate changes include long-term changes in precipitation patterns and the pattern or severity of winds and storms.

The Earth receives the vast majority of its energy from the sun. The energy that comes from the Earth's core contributes less than 0.03% of the Earth's surface energy, and other effects are even smaller. Global surface temperatures are determined largely by the amount of solar radiation that reaches the Earth and is retained on the surface and in the lower atmosphere. The major factors determining global surface temperatures are shown in Table 3.1.

Table 3.1: Major factors determining global surface temperatures

Factor	Description	Moderating Effects
Global Insolation	Amount of solar radiation reaching the top of the Earth's atmosphere	Solar Cycles, Milankovitch Effect
Earth's Albedo	Reflection of solar radiation back into space from the atmosphere and Earth's surface	Clouds, Aerosols, Particulates, Volcanic Activity, Surface Ice and Snow
Surface Absorption	Surface absorption of solar energy	Ocean Effect, Vegetation Effect
Greenhouse Effect	Absorption and re-radiation by the atmosphere of solar energy and heat from Earth's surface	Atmospheric concentrations of water vapor, carbon dioxide, ozone, methane, nitrous oxide, and chlorofluorocarbons

Global Insolation

Global insolation, when the Earth is one astronomical unit (92.96 million miles) from the sun, is 1360.8 ± 0.5 W/m². The amount of solar irradiance varies over time in cycles related to fluctuation in the magnetic activity of the sun (Figure 3.1, blue dotted line).[57] The solar cycle, or solar magnetic activity cycle, is a periodic 11-year change in the sun's activity and appearance, including changes in the amount of solar radiation and ejection of solar material, number and size of sunspots, flares, and other manifestations. Total solar irradiance (TSI) patterns have changed very little in the last 70 years. TSI has certainly not increased over this time period. In fact, since 1950 the 11-year average TSI has decreased.

Figure 3.1: Solar cycles from 1880 to 2018. Note fluctuation in yearly total solar irradiance as indicated in the blue-dotted line.

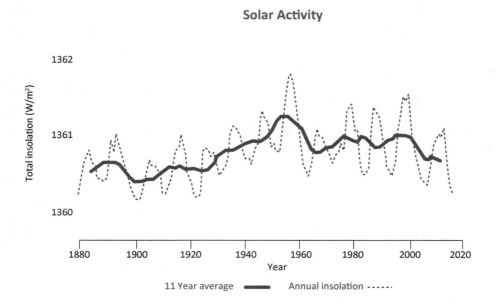

The energy coming from the sun is primarily in the visible or ultraviolet electromagnetic spectrum, and energy that is absorbed and reradiated by the Earth's surface is primarily in the infrared spectrum.

The Milankovitch Effect

The movement of the Earth around the sun is not a simple circular orbit. The movement is elliptical, meaning the Earth is not always exactly one astronomical unit from the sun. Furthermore, as the Earth moves around the sun, it gradually changes its tilt toward the sun and wobbles on its axis. These orbital movements are due to the gravitational effects of the sun, moon, and other planets and are predictable.

Milutin Milankovitch studied these orbital movements and related them to the amount

---

and distribution of solar radiation.[58] He hypothesized that changes in the Earth's elliptical orbit around the sun (eccentricity), the angle of the Earth's axis toward the sun (obliquity), and the wobble of the Earth's axis (axial precession) could be used to calculate changes in solar radiation in the northern and southern hemispheres. These changes in solar irradiance could then be used to calculate an orbital forcing effect that could be used to explain (or predict) global surface temperatures and climate.

Radiative forcing is a change in the amount of radiant energy retained by the Earth, including the Earth's atmosphere. It is the difference between the amount of incoming energy from the sun and outgoing radiation from the Earth and can be positive, negative, or in equilibrium. There are a variety of factors that can change the amount of radiative forcing, including changes in the Earth's orbit (orbital forcing) or atmosphere (atmospheric forcing) and solar irradiance (solar forcing). For example, the Earth's orbit gradually changes from a circle to an ellipse over a period of ~100,000 years. When the orbit is an ellipse, solar radiation is higher when the Earth is closer to the sun and lower when it is farther away. The obliquity of the Earth gradually changes from 22.1° to 24.5° in periods lasting 41,000 years. The current tilt is 23.5° and gradually decreasing. When the tilt is larger, the seasons become more extreme.

Currently, the Earth's orbit around the sun is nearly a circle (eccentricity of 0.0167) and average annual solar surface radiation at 65°N latitude is 480 watts/m²/day, which is very consistent with the level of insolation predicted by Milankovitch. The Milankovitch hypothesis has been further tested using paleoclimate data, including ice cores, deep ocean cores, and deep cores drilled in rock. These data can be used to estimate global surface temperatures from as long ago as 250 million years. As can be seen in Figure 3.2, the fit of solar irradiance at 65°N in July with global surface temperature anomalies predicted from the Vostok ice core is good but not perfect.[59] As a result of this and other analyses, there are a number of unanswered scientific questions regarding the relative contribution of the various orbital effects and the degree to which they interact.

The paleoclimate data in Figure 3.2 show that the variation in Earth's temperature is less extreme (i.e., less variable) than expected from the variation in the intensity of solar radiation. The last ice age began about 115,000 years ago and ended about 12,000 years ago. The Earth's current orbit is nearly circular and is in a period of moderate negative orbital forcing (see figure 3.2, blue line), which should lead to cooling of the Earth's surface and the next ice age. Recently, the interaction of orbital forcing with forcing

---

58   M. Milankovitch, *Mathematische Klimalehre und Astronomische Theorie der Klimaschwankungen.* Handbuch der Klimatologie series. (Berlin: Gebrüder Borntraeger, 1930).

59   http://www.climatedata.info/proxies/ice-cores/

due to the concentration of greenhouse gases in the atmosphere has been studied. [60] This research suggests that the Earth narrowly escaped a period of significant cooling and glaciation just prior to the industrial revolution and that, under current conditions, the Earth is expected to remain in an interglacial period, without large-scale glaciation or complete deglaciation, for a very long time.

Figure 3.2: Milankovitch Cycles and Temperatures from the Vostok Ice-core.

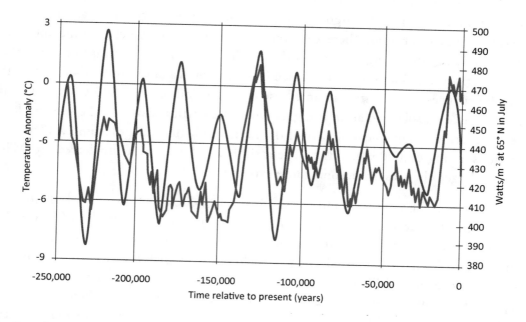

The Earth's Albedo

The Earth's atmosphere and surface reflect some of the incoming solar radiation back into space. The degree to which the surface and atmosphere are reflective is called the Earth's "albedo." A surface that is a perfect reflector has an albedo of 1.0. Ice and snow have an albedo of about 0.84, indicating that ice and snow reflect about 84% of incoming solar radiation. Dark vegetation, on the other hand, has an albedo of about 0.14, indicating dark vegetation absorbs most solar radiation and only reflects about 14%. Satellites are able to measure the amount of reflected solar radiation. Over the past decade or more, the Earth's albedo has been 0.29 ± 0.01 with some year-to-year fluctuation, indicating the Earth reflects 29% ± 1% of incoming solar radiation back into space. [61] (Figures 3.3 and 3.4)

---

60   A. Ganopolski et al., "Critical Insolation-CO_2 Relation for Diagnosing Past and Future Glacial Inception," *Nature* 529 (2016): 200-203.

61   E. Palle et al., "Earth's Albedo Variations 1998-2014 as Measured from Ground-Based Earthshine Observations," *Geophys Res Lett*. 43(9) (2016): 4531-4538.

Figure 3.3: The Earth's albedo from March 2000 to 2013. Blue areas have a low albedo; yellow, pink, and red are areas of progressively higher albedo. Gray indicates no data are available.

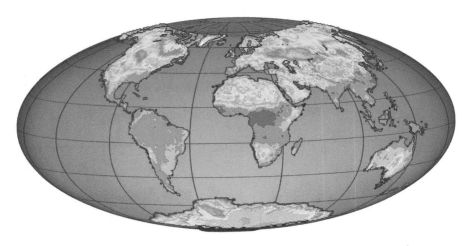

Figure 3.4: The Earth's albedo anomaly March 2000 to present using January 2000 as the reference value[62].

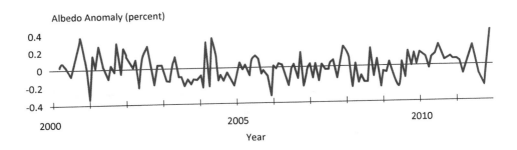

No significant long-term trend in the Earth's albedo has been detected. Nevertheless, the magnitude of the albedo is very important in determining the net amount of solar energy retained by the Earth (i.e., the magnitude of radiative forcing).[63] Major changes in clouds, aerosols, particulates, volcanic activity, and surface ice can have a large effect. It has been estimated that lowering the albedo from 0.29 to 0.20 through melting of polar ice caps would have the same warming effect as doubling the concentration of carbon dioxide in the atmosphere.

Sulfate aerosols are naturally present in the Earth's stratosphere and increase during some volcanic eruptions. The particles consist of a mixture of tiny sulfuric acid and water droplets that reflect incoming solar radiation and transiently increase Earth's albedo. After

62   https://Earthobservatory.nasa.gov/images/84499/measuring-Earths-albedo.

63   G. L. Stephens, D. O'Brien, P. L. Webster, P. Pilewski, S. Kato, J.-I. Li, "The Albedo of Earth," *Rev. Geophys.* 53 (2015): 141-163, doi: 10.1002/2014RG000449.

a few years, the particles degrade, and their effect dissipates. The effect of recent volcanic eruptions on transmission of solar radiation through the stratosphere is shown in Figure 3.5.[64] Sulfate aerosols are also found in the lower atmosphere as a result of burning coal, but these aerosols do not significantly increase Earth's albedo.

Figure 3.5: Solar radiation transmission through the Earth's stratosphere as measured at the Mauna Loa Observatory in Hawaii. [Note: A reduction in solar radiation transmission reflects an increase in the Earth's albedo.]

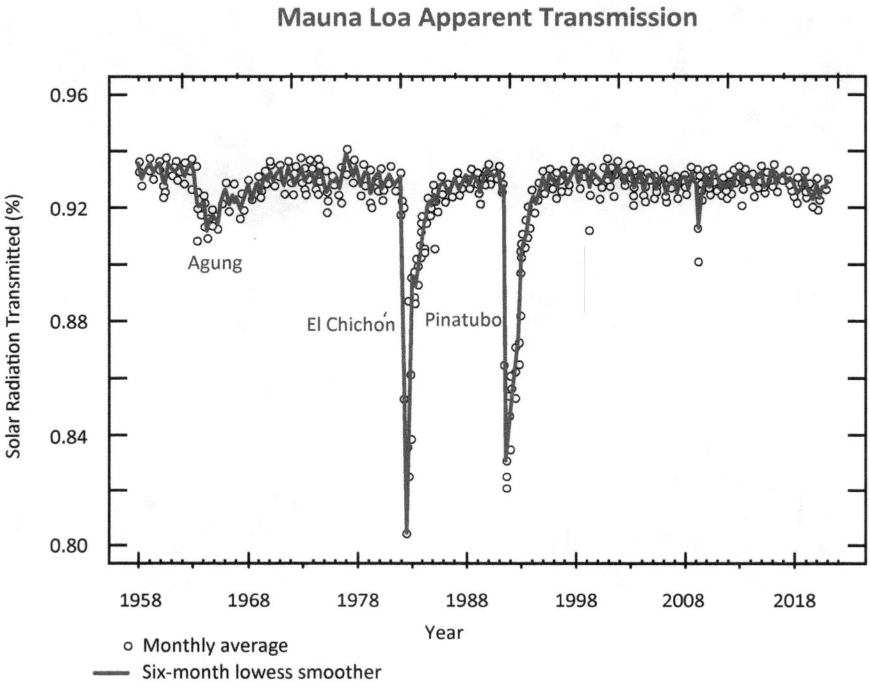

Mauna Loa Apparent Transmission

Atmospheric Absorption

The Earth receives electromagnetic radiation from the sun. The spectrum that reaches the top of the Earth's atmosphere has a wavelength from 100 nanometers (nm) to >1 millimeter (mm), as shown in Figure 3.6.[65] Ultraviolet radiation has a wavelength of 100-400 nm, visible light from 400-700 nm, and infrared from 700 nm to 1 mm.

Earth's atmosphere contains gases that absorb some of the incoming solar radiation and some of the infrared radiation emitted from the surface. For example, ozone in the Earth's upper atmosphere absorbs greater than 95% of all incoming ultraviolet (UV) radiation. Thus, the ozone layer protects us from UV light from the sun that can cause damaging

64   https://commons.wikimedia.org/wiki/File:Mauna_Loa_atmospheric_transmission.png.

65   https://www.visionlearning.com/en/library/Earth-Science/6/
      Factors-that-Control-Earths-Temperature/234

effects on our skin. Other atmospheric gases, such as water vapor, carbon dioxide, and methane, absorb some of the incoming infrared radiation, which we sense as heat. The atmosphere absorbs very little solar radiation in the visible light spectrum (380 to 740 nm), photosynthetically active radiation (400 to 700 nm), most radio waves (5.0 cm to 10 m), and some other wavelength "windows." The net effect is that the atmosphere absorbs about 22.6% ± 0.1% of incoming solar radiation.

Figure 3.6: Electromagnetic spectrum at the top of the Earth's atmosphere and at sea level

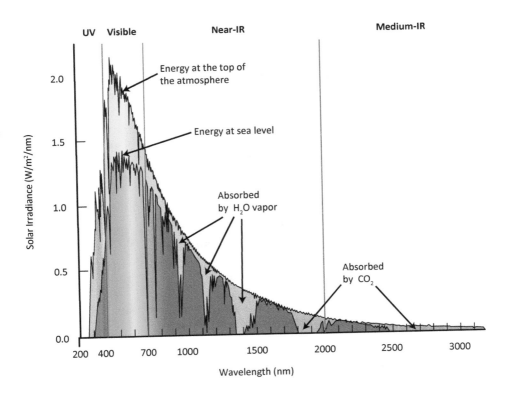

Surface Absorption

The amount of solar radiation reaching the surface of the Earth is 54.7% ± 0.03% of the total solar radiation reaching the top of the atmosphere. About 11.7% of the solar radiation reaching the surface is reflected back into space and is accounted for in the Earth's albedo. Ice sheets, snow, glaciers, clouds, and sand have the highest albedo and reflect much of the incoming solar radiation reaching these surfaces. Oceans and vegetation have a low albedo and absorb most of the remaining 43.0% ± 0.03% reaching the Earth's surface. Oceans absorb most of this energy and play a large role in moderating the surface temperature of Earth.

The amount of energy oceans can absorb is a function of their heat capacity. The heat capacity of a substance is the amount of energy required to heat one unit of the substance one degree Kelvin (or one degree Celsius) under specified pressure conditions. The units of heat capacity can be expressed as mass (e.g., grams), volume (e.g., cubic centimeters), or other standard measures (e.g., moles). The heat capacity of water is 4.18 joules/gm./°C, for air it is 1.16 J/gm./°C, and for soil or sand it is about 0.8 J/gm./°C. Thus, on a weight basis, the heat capacity of water is about 5 times greater than land and about 4 times greater than air. In reality, the relationship between heat transfer and surface temperature change is more complex, because the oceans and the atmosphere are not homogeneous. Most of the energy in the oceans is in the top 75 meters. The temperature of the deep ocean is 0-4°C compared with much warmer sea-level temperatures. Pressure increases as depth increases, and the amount of energy needed to raise the temperature goes up. Ocean currents in the North Atlantic and southern oceans allow for heat transfer from the surface to deeper waters, but this process is slow, and equilibration takes decades or centuries.

The net heat absorbed by the oceans is not uniformly distributed across the globe. The tropical, central, and eastern Pacific Ocean has a temperature oscillation called the El Niño Southern Oscillation (ENSO) that has a major effect on weather in countries bordering the Pacific and, possibly, the Earth as a whole. ENSO consists of warm water cycles called El Niño and cold water cycles called La Niña. El Niño is a pooling of warm water in the Pacific Ocean that heats the atmosphere, and La Niña is an upwelling of cold water that cools the atmosphere.[66] (Figure 3.7) Currently, an El Niño occurs every 2 to 7 years. These oscillations also affect atmospheric pressure, cloud patterns, precipitation, and storm patterns over the Pacific, the Western Hemisphere, Australia, and Asia.

Figure 3.7: El Niño oscillation in 2016. Red represents warm water and blue represents cold water.

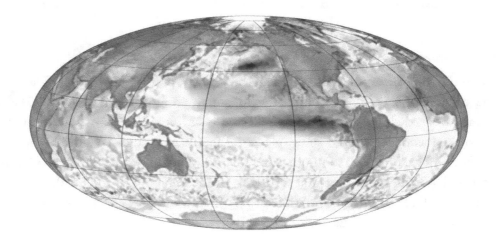

---

66   https://blogs.ei.columbia.edu/2016/02/02/el-nino-and-global-warming-whats-the-connection/.

The relationship of ENSO with global climate change is not a scientifically settled matter. It is known that an El Niño cycle will contribute to an increase in global temperatures, but it is not clear if rising global and ocean temperatures will intensify ENSO cycles. Recently, rapid warming of the tropical oceans during 1981 to 2018 has been reported, leading to major changes in the atmospheric Madden-Julian Oscillation (MJO). The MJO has far-reaching effects on the Earth's climate system, including ENSO, tropical cyclones, and monsoons. Observed changes in the MJO are expected to increase rainfall over Southeast Asia, Northern Australia, Southwest Africa, and the Amazon, and decrease rainfall over the west coast of the US.

Greenhouse Gases:

The surface of the Earth radiates energy previously absorbed from the sun in the infrared electromagnetic spectrum, which is perceived as heat. The outgoing radiated heat does not pass through the atmosphere nearly as well as the incoming solar radiation because gases in the atmosphere absorb it and reradiate much of it back toward the surface. These gases are called "greenhouse gases" because they work just like glass in a greenhouse or an invisible blanket around the Earth, capturing and reradiating heat. Without this blanket of gases, Earth's average surface temperature would be much lower, about -18°C by some estimates, and would be incompatible with human life. Thus, atmospheric greenhouse gases are essential to maintaining a habitable environment for humans.

The principle greenhouse gases are water vapor, carbon dioxide, methane, nitrous oxide, and chlorofluorocarbons. Each of these atmospheric gases varies in concentration, residence time in the atmosphere, and in global warming potential. (Table 3.2) Each gas has a different effect and residence time in the atmosphere. In order to add these effects together, a common unit is needed. The common unit generally used is the "CO_2-equivalent" or the "Global Warming Potential." The CO_2-equivalent concentration is derived by calculating the concentration of CO_2 that would be required to produce the same radiative forcing effect in watts/m^2 produced by the greenhouse gas in question. The Global Warming Potential (GWP) is the number of tons of carbon dioxide that would need to be emitted to produce the same global warming effect as one ton of the greenhouse gas. The GWP depends on the timeframe evaluated. Since methane has a shorter residence time in the atmosphere than carbon dioxide, the GWP for methane is higher for shorter periods than longer periods. The 100-year GWP is used for most purposes.

Table 3.2: Atmospheric concentration, residence time, and Global Warming Potential for common greenhouse gases

Gas	Concentration* In Dec. 2018	Residence time (Years)	100-year Global Warming Potential	20-year Global Warming Potential
Carbon dioxide	410 ppm	20-200	1	1
Methane	1868 ppb	~12	28-36	84-87
Nitrous Oxide	332 ppb	~114	265-298	268

*ppm = parts per million = μmoles/mole; ppb = parts per billion = nanomoles/mole

Water vapor is a potent greenhouse gas, but its concentration in the atmosphere is a direct result of global surface temperatures, so it differs from the other greenhouse gases. About 50% of the greenhouse effect is due to water vapor in the atmosphere, 20% is due to carbon dioxide, and 25% is due to clouds. The remaining 5% is due to methane, nitrous oxide, and chlorofluorocarbons. Over time, increased concentrations of greenhouse gases will lead to a rise in global mean surface temperatures unless there are changes in other components of the Earth's energy budget that offset the greenhouse effect. For example, particulates and aerosols in the atmosphere produced by volcanoes or dust storms can reflect incoming solar radiation and transiently cool the Earth's surface.

The Earth's Energy Budget

The Earth's energy budget is the balance of incoming and outgoing energy.[67] (Figure 3.8) It is the sum of various energy fluxes that describe the distribution of incoming radiation from the sun and the sources of outgoing radiation. The incoming radiation is short wave (wavelength less than 700 nm), mostly in the visible or ultraviolet range, while the outgoing radiation is long-wave (wavelength greater than 700 nm) in the infrared range. The incoming radiation can produce heat or work (e.g., photosynthesis) in the various energy reservoirs. The largest reservoirs of energy on Earth are the oceans. NASA's CERES (Clouds and the Earth's Radiant Energy System) Program and the ARGO Program have studied the Earth's energy balance. The ARGO Program is named after the mythical Greek ship "Argo," because the array of floats works in partnership with the Jason Earth Observing Satellites that measure sea levels.

Radiative forcing is the net amount of solar energy retained by the Earth, measured in watts/m², and is a key parameter in climate science. The oceans absorb the majority of this energy, but some energy warms the land and lower atmosphere (troposphere). The

---

67   https://en.wikipedia.org/wiki/Earth%27s_energy_budget showing_the_Earth%27s_
       energy_budget,_which_includes_the_greenhouse_effect_(NASA).png.

relative contribution of the factors influencing radiative forcing is shown in Figure 3.9.[68] The net anthropogenic component is currently estimated to be 2.0 ± 0.9 W/m². The amount of radiative forcing can vary from year-to-year based on changes in any of the factors contributing to net energy retention. For example, a large volcanic eruption can cause a net reduction in radiative forcing due to an increase in the Earth's albedo, while rising levels of greenhouse gases can cause an increase.

Figure 3.8: The Earth's energy budget.

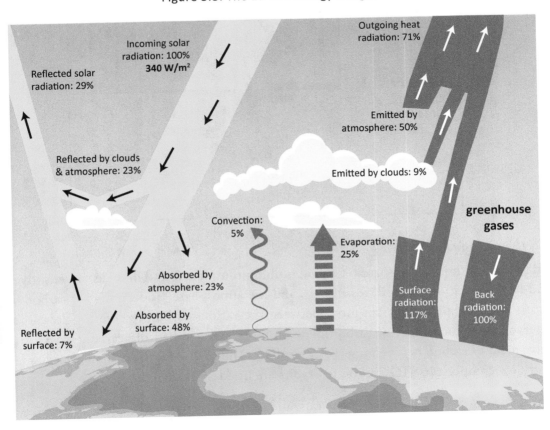

___
68   https://commons.wikimedia.org/wiki/File:Radiative-forcings.svg.

Figure 3.9. Radiative-forcing components

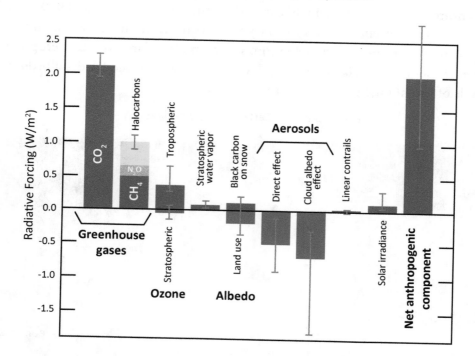

Earth's Carbon Cycle

All life on Earth is carbon-based. Carbon is ubiquitous; it is found in rocks, oceans, rivers and lakes, living things, soil, fossil fuels, and the atmosphere. The amount of carbon on Earth is fixed, but carbon can move between the various carbon reservoirs. The carbon reservoirs and movement of carbon between reservoirs is called the carbon cycle.[69] The movement of carbon between reservoirs is called a "flux." The Earth's carbon cycle has fast and slow components. The carbon reservoirs are listed in Table 3.3.

Through a series of chemical reactions and tectonic activity, the slow carbon cycle takes between 100-200 million years to move carbon between rocks, soil, ocean, and atmosphere. Rain and chemical weathering of rock is an important part of the slow process, as is formation of calcium carbonate in the ocean, formation of fossil fuels, and volcanic activity due to movement in the Earth's tectonic plates. The fast carbon cycle is diagramed in Figure 3.10.[70]

Fluxes in this cycle act over years or decades and contribute to the greenhouse effect. Table 3.4 summarizes the fluxes and includes the effect of the flux on the atmosphere. About 10% of the carbon in the fast cycle moves between reservoirs each year.

---

69   H. Riebeek and R. Simmon, "The Carbon Cycle" (2011), www.Earthobservatory.nasa.gov.
70   https://Earthobservatory.nasa.gov/features/CarbonCycle.

Table 3.3: Earth's carbon reservoirs

Earth Carbon Reservoir	Carbon Content (Gigatonnes)	Major Deposits	Cycle
Mantle	10^5 to 10^8	Limestone, Shale, Fossil fuels	Slow
Deep Ocean	38,000	Inorganic carbon	Slow
Ocean Surface	1,000	Biomass, Calcium carbonate, Coral	Fast
Soil biomass	1,500	Decaying biomass, microorganisms	Fast
Atmosphere	750 to 800	CO_2, methane	Fast
Plants	560	Trees, plants, grasses	Fast
Total Slow Reservoirs	10^5 to 10^8	Inorganic carbon	Slow
Total Fast Reservoirs	3,800 to 4,000	Organic carbon	Fast

The fast carbon cycle involves the movement of carbon between biological systems or into and out of the atmosphere and oceans. In 2019 humans emitted about 36.8 billion tonnes of carbon dioxide or 10 gigatonnes of carbon into the atmosphere. About 4.5-5.0 gigatonnes is retained in the atmosphere, and the remainder produces biomass through photosynthesis or is taken up by the oceans. The carbon retained in the atmosphere is mostly in the form of carbon dioxide.

Data Sources and Methods

Climate scientists measure climatic conditions on land, in oceans, and in the air. Land-based stations, ships, buoys, and satellites collect the data and are located around the world. The number of these stations differs between monitoring organizations. The Japanese Meteorological Administration covers about 85% of the world and the US National Aeronautical and Space Administration covers about 99% of the world with about 7,000 stations. Land and ocean measurements have been made since 1850 by the United Kingdom, while the US has been collecting data since 1880. In 1979 orbiting satellites began collecting data, including measurements of atmospheric temperatures.

Figure 3.10: The Earth's fast carbon cycle in gigatonnes. Black numbers are natural fluxes, red are human contributions, and white are stored carbon.

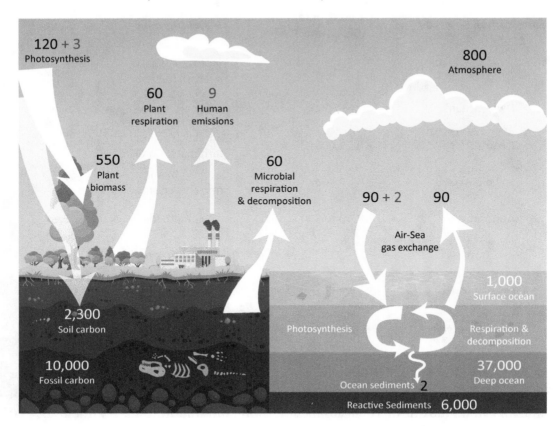

Over the decades, the sites, instrumentation, protocols, and statistical methods for gathering, analyzing, and reporting climate data have changed, as have the scientific organizations that measure and report the data. There are now four major organizations that report climate data: *HadCRUT4* produced by the Hadley Center of the UK Met Office and the Climatic Research Unit of the University of East Anglia, *GISTEMP* produced by the US National Aeronautical and Space Administration (NASA), *MLOST* produced by the US National Oceanic and Atmospheric Administration (NOAA), and *JMA* produced by the Japanese Meteorological Administration. Each of these organizations collects daily measurements that are compared with a long-term average for the site (e.g., 30-year average), and a daily anomaly is recorded. Daily anomalies are averaged for the month and for the year for each site and then combined to produce a global average, which is called "the global average surface temperature anomaly." The data from the northern and southern hemispheres are equally weighted.

Table 3.4: Fast carbon cycle fluxes per year and the effect of
the flux on atmospheric carbon concentrations

Carbon Flux	Amount (Gigatonnes per year)	Effect on Atmosphere	Comment
Photosynthesis	123	Removal	Land use change can result in long-term decrease
Plant, human, and animal respiration	60	Addition	Land use change can result in long-term decrease
Soil; microbial respiration and decomposition	60	Addition	Melting of the tundra could increase the amount
Fossil Fuel Burning	9.9	Addition	2018 Carbon Emissions
Net Land Effect	6.9	Addition	
Surface Ocean Uptake	92	Removal	Likely to increase as atmospheric carbon concentration increases
Surface Ocean Emissions	90	Addition	
Deep Ocean Uptake	2	Removal	
Net Ocean Effect	2	Removal	
Atmosphere	4.9	Net addition	Equates to ~3 ppm CO_2-equivalent increase/year

The instruments and methods used are subject to error, and there are some areas of the Earth that are not covered, especially in the polar regions. Over time, how, when, and where the temperatures are recorded has changed, necessitating adjustment or correction of the data.[71] For example, adjustments have been made for the type of thermometer used, the location of the weather station, the time of day measurements were taken, the proximity of the weather station to cities, and shading of the equipment. The methods used for adjustment differ between reporting organizations, so there are small differences in the average global anomalies reported. The aggregate effect of these adjustments is small (less than10%), especially for the past 50 years. The largest difference between the

---

71   Z. Hausfather, "How Data Adjustments Affect Global Temperature Records" (2017), https://www. carbonbrief.org/explainer-how-data-adjustments-affect-global-temperature-records.

raw data and the adjusted data occurs for those anomalies reported before 1940. Prior to 1940, adjustment results in an increase in the global mean surface temperature anomaly, which reduces the estimated magnitude of global warming since 1880.

Paleoclimatology is the study of ancient climates. Paleoclimatologists use surrogate markers to estimate and reconstruct climatic conditions prior to the modern era (1880 to present). These surrogate markers include tree rings, coral growth, ice core samples, borehole samples, isotope measurements, and changes in glaciers. Because these estimates are from indirect evidence and are much more sparse than modern direct measurement, there is much more uncertainty surrounding the data. By using multiple data sources, other indirect corroborating evidence, and statistical methods, paleoclimatologists try to minimize the uncertainty but are careful to account for it in any inferences they draw from their data.

The major sources of data on the Earth's energy budget are from satellites and from the Earth's oceans. NASA's Clouds and the Earth's Radiant Energy System (CERES) Program is a part of NASA's Earth Observing System (EOS). The first CERES instrument was launched in 1997 and the most recent in 2017. There are now six CERES instruments in orbit providing scientists with critical data on Earth's energy budget, including data on reflected solar radiation and thermal radiation from Earth. The CERES program issues public reports on the quality and accuracy of the data on Earth's energy budget. Potential sources of error in the system include the detection limits of the instruments, the calibration of the detectors, the algorithms used to calculate energy fluxes from the raw data, and other sources. The limit of detection varies by instrument and type of radiation and is between 1 and 5 W/m^2. This represents less than 2% of the fluxes being measured.

ARGO was established in 2000 and is an international program to monitor the Earth's oceans. ARGO has nearly 4000 drifting floats throughout the world's oceans. To maintain the ARGO array, about 800 floats are needed per year. These floats monitor currents, temperature, water density, salinity, and bio-optical properties of the oceans from the surface to a depth of 2000 meters. Measurements are taken in a 10-day cycle for each float and are transmitted by satellite to centers in Australia, France, Italy, Japan, Norway, the UK, and the US for analysis. A real-time data delivery and quality control system delivers 90% of data profiles to users via two global data centers within 24 hours. A delayed mode, more comprehensive quality control and data review system has also been established. ARGO works with universities, government labs, and meteorological centers. There are several ARGO enhancements that are in various stages of development and implementation. These include extended coverage to the ocean bottom, additional floats equipped with bio-geochemical sensors, and enhanced spatial coverage in boundary current regions and equatorial regions.

Climatologists have developed mathematical formulas or mathematical "models" that relate global surface temperatures on a large scale to the factors in Table 3.1: global insolation, the Earth's albedo, energy absorbed by the Earth's surface, and the effects of the atmosphere, especially the effect of greenhouse gases.[72] Climate models incorporate a number of parameters whose values are not always precisely defined by observational studies. In these cases, a range of values can be assessed and an analysis undertaken to define the range of possible effects.

The details and methods used in building mathematical models are important. Validation with data not used to generate the model is a critical step. There are many climate models and most perform well in predicting past global surface temperatures. During the past three decades many of these models have been refined and applied prospectively. Some have performed well in predicting changes in global surface temperatures. They have also worked well when major atmospheric events such as large volcanic eruptions have occurred. The models perform less well in predicting precipitation or other climatic changes, especially on a regional basis.

Current Data

The Earth has been warming for the last 100 years (Figure 3.11).[73] From 1880 to 1910 the global mean surface temperature (GMST) is estimated to have been 13.9 ± 0.2°C. During this early industrial period world energy consumption was less than 3% of its current level. During the 20th century, the GMST is estimated to have been 14.0 ± 0.2°C. In the post-World War II period from 1951 to 1980, during which global industrialization grew rapidly, the GMST is estimated to have been 14.2 ± 0.01°C. The GMST for 2019 was 14.98°C, having increased by 0.98°C (~1.76°F) relative to the 20th-century mean and 1.08°C since the early industrial period.[74] Since 1970 the rate of global temperature rise has been approximately 0.02°C/year, and for the past 10 years the rate of increase has been about 0.03°C/year. According to NOAA, GMST in 2020 was the second warmest year on record behind 2016 and was 1.2°C higher than the average from 1951-1980, despite a cooling La Niña effect from the El Niño Southern Oscillation.

---

72   D. A. Randall, R. A. Wood, S. Bony, R. Colman, T. Fichefet, J. Fyfe, V. Kattsov, A. Pitman, J. Shukla, J. Srinivasan, R. J. Stouffer, A. Sumi, and K. E. Taylor, "Climate Models and Their Evaluation," *Climate Change 2007: The Physical Science Basis*, contribution of Working Group I to the Fourth Assessment Report of the Intergovernmental Panel on Climate Change, eds. S. Solomon, D. Qin, M. Manning, Z. Chen, M. Marquis, K. B. Averyt, M. Tignor, and H. L. Miller (Cambridge University Press, Cambridge, United Kingdom and New York, NY, USA: 2007).

73   https://data.giss.nasa.gov/gistemp/graphs/.

74   R. Lindsey and L. Dahlman, "Climate Change: Global Temperature" (2020), https://Climate.gov.

Figure 3.11: Adjusted global mean surface temperature anomaly (GMST) using the average for the 30-year period from 1951-1980 (14°C) as the baseline. Smoothing was done by Locally Weighted Scatterplot Smoothing (LOWESS). Standard error bars are shown in blue.

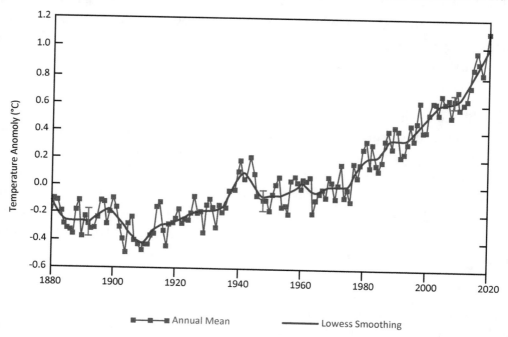

The rise in global mean surface temperature is not uniformly distributed across the surface of the Earth. (Figure 3.12) The northern hemisphere is most effected while some other regions have actually cooled. All of the cooler regions are over oceans due to the heat absorption capacity of the oceans and other ocean effects, such as currents that can bring cooler water from ocean depths to the surface. Most of the warmest regions are over land in the northern hemisphere. The Land-Ocean Temperature Index (L-OTI) is the GMST anomaly relative to the average GMST from 1951-1980. The estimate for the global increase in average Land-Ocean Temperature Index (L-OTI) from January 2014 to December 2018 compared to the average from 1951 to 1980 is 0.86°C.

The average L-OTI from 2014 to 2018 for each latitude zone, or the zonal mean L-OTI, is plotted against latitude in Figure 3.13. Northern latitudes have warmed more than southern latitudes, and both polar regions are warming, although the north polar region is warming much faster than the south polar region. The most northern latitudes (zones) have warmed more than 2.5°C, illustrating the uneven geographic distribution of global warming.

Figure 3.12. Average L-OTI anomaly for 2014-2018 relative to 1951-1980

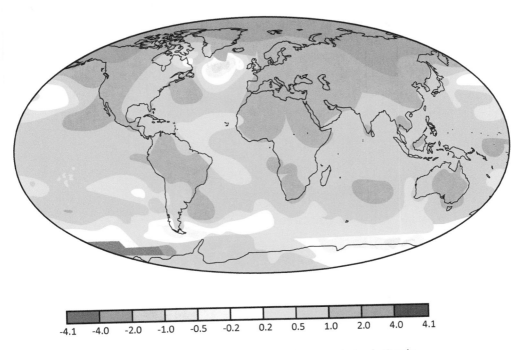

Figure 3.13. Zonal mean L-OTI anomaly by latitude

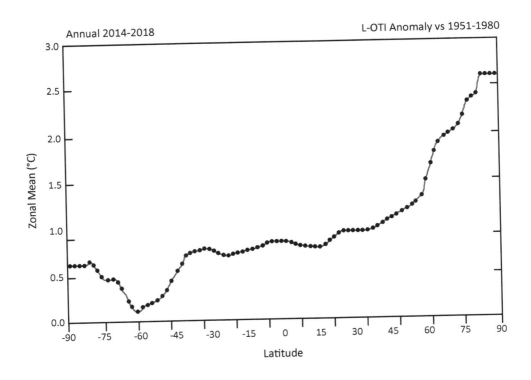

Paleoclimatologists have estimated global surface temperatures for the last 11,300 years (Figure 3.14).[75] The estimated global mean surface temperatures have fluctuated between ±0.5°C from the global mean surface temperature for the period from 1961-1990. As can be seen, the magnitude and rate of rise in global mean surface temperatures during the last 100 years is unprecedented in the last 11,300 years.

Figure 3.14: Estimated global surface temperatures for the last 11,300 years

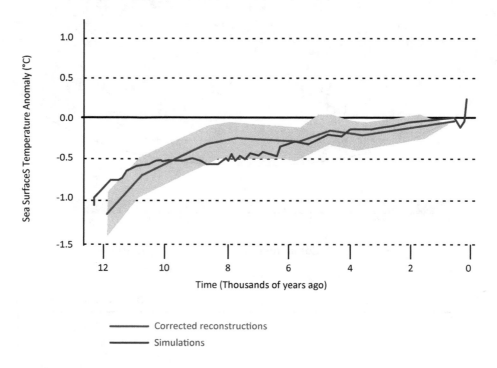

Changes in solar irradiance, atmospheric ozone concentrations, volcanic activity, and cloud cover cannot explain the rise in global mean surface temperatures over the past 100 years. Changes in these factors during the 20th century would actually have resulted in a small amount of cooling since 1960. As seen in Figure 3.15, solar irradiance has not increased in the past 50 years, while surface temperatures have risen significantly. Thus, sunspot activity, solar cycles, and the Milankovitch effect cannot explain the rise in surface temperatures. The increase in global temperatures is not due to "normal" cyclical changes in the amount of solar radiation reaching Earth.

---

75   S. A. Marcott, J. D. Shakun, P. U. Clark, and A. C. Mix, "A Reconstruction of Regional and Global Temperature for the Past 11,300 Years," *Science* 339 (2013): 1198-1201, doi: 10.1126/science.1228026.

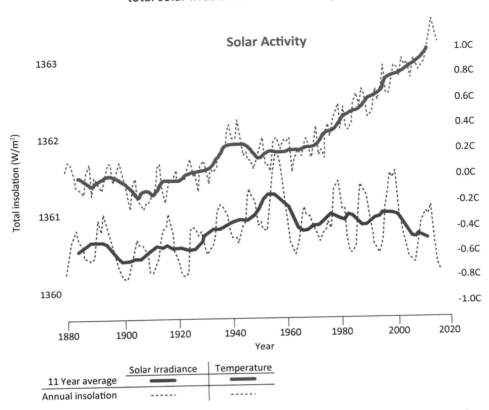

Figure 3.15: Global mean surface temperature anomaly and total solar irradiance from 1880 to present[76]

Land has warmed more than oceans.[77] (Figure 3.16) The northern hemisphere has warmed more than the southern hemisphere because of northward cross-equatorial ocean heat transport in currents and other factors.

Volcanic eruptions can eject large quantities of gases, sulfate aerosols, and ash into the atmosphere. On average, volcanic eruptions eject about 0.6 ± 0.2 billion tonnes of carbon dioxide into the atmosphere each year. This amount is about 1.5% of the CO_2 emitted from anthropogenic sources. Individual eruptions generally emit large quantities of CO_2 for only a few hours or days, so the rise in global atmospheric CO_2 concentrations over the last 50 years cannot be explained by volcanic eruptions. However, volcanoes can also emit very large quantities of ash and other particulates into the stratosphere that can persist for long periods. These particulates can increase the Earth's albedo and cause a cooling effect on the troposphere and a warming effect on the stratosphere that can last for many months or years, depending on the characteristics of the eruption. (Figure 3.17)

76   https://climate.nasa.gov/blog/2910/what-is-the-suns-role-in-climate-change/

77   https://data.giss.nasa.gov/gistemp/graphs_v3/.

Figure 3.16: Air temperature anomalies over land and over ocean since 1880. Temperature anomalies are relative to the average surface temperature from 1880 to 1900.

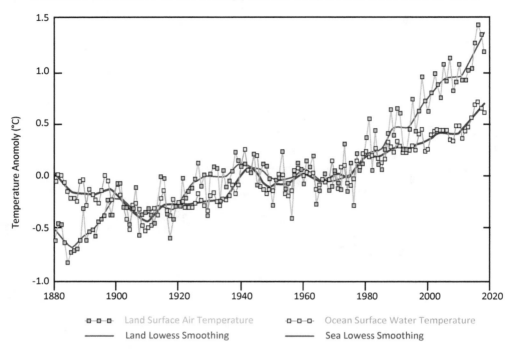

Figure 3.17: Average stratosphere and troposphere global temperature anomalies related to volcanic activity.[78] The temperature anomaly is relative to the 1951-1980 global mean surface temperatures. Note that generally the lower stratosphere warms and the troposphere cools after an eruption.

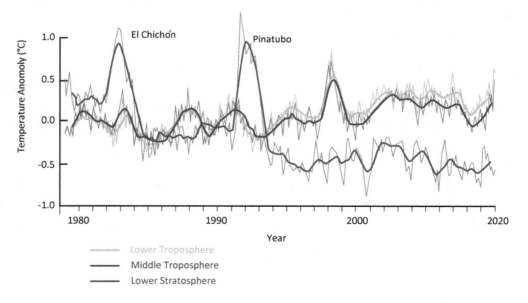

78   https://Earthobservatory.nasa.gov/features/GlobalWarming/page4.php.

Carbon Dioxide

The atmospheric carbon dioxide concentration at the Mauna Loa observatory in April 2020 was 416.2 parts per million (ppm), an increase of 2.88 ppm from the prior year. Atmospheric carbon dioxide concentrations have increased by over 60% during the last 60 years. (Figure 3.18) This increase cannot be explained by volcanic eruptions, release of CO_2 from the oceans, or production of CO_2 by human, animal, or plant respiration.

Figure 3.18: Atmospheric carbon dioxide concentrations at Mauna Loa, Hawaii

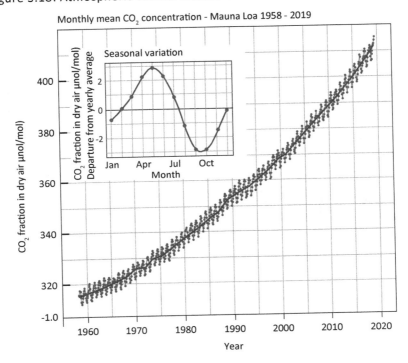

For all of human history (i.e., ~300,000 years), atmospheric carbon dioxide concentrations have been 180 to 280 ppm and average surface temperatures have been 13°C to 14°C (55 to 57°F). Changes in atmospheric carbon dioxide concentrations persist for 100 to 200 years, depending on the dynamics of the fast carbon cycle. Currently, annual world CO_2 emissions are about 34-36 billion tonnes and are increasing by about 350-700 million tonnes per year. Over the next 40 years, atmospheric carbon dioxide concentrations will likely increase by 90-120 ppm or more, depending on the effectiveness of future carbon mitigation strategies. The annual amount of carbon dioxide emitted globally is likely to exceed 44 billion tonnes by 2050 unless measures to control, capture, store, or utilize carbon emissions are developed and deployed on a large scale. Atmospheric carbon dioxide concentrations in 2050 will likely be between 480 ppm and 520 ppm.

Paleoclimatologists have used the data obtained from ice cores to study the relationship between surface temperatures, atmospheric carbon dioxide concentrations, and atmospheric particulates from volcanoes over the last 800,000 years. For example, the ratio of two isotopes of oxygen, O^{16} and O^{18}, can be used to estimate past temperatures based on the fact that O^{18} is heavier than O^{16}.[79] Because O^{16} is lighter, it evaporates more quickly from water when temperatures are high. By measuring the ratio of O^{18}/O^{16} in ice core samples, estimates of past Antarctic surface temperatures can be obtained. In addition, atmospheric carbon dioxide concentrations can be measured in trapped air bubbles, as can the concentration of dust particles. Data from Dome C in Antarctica is shown in Figure 3.19.[80] Increases in atmospheric carbon dioxide concentrations are highly correlated with increases in surface temperature. Cooling is correlated with high levels of dust particles that cause an increase in the Earth's albedo.[81]

Figure 3.19. Changes in Antarctic surface temperatures, atmospheric carbon dioxide concentrations, and dust flux over the last 800,000 years. The data are obtained from the Dome C ice core in East Antarctica. BP means "before present." Please note the time scale: the origin is the present; 800 refers to 800,000 years before the present.

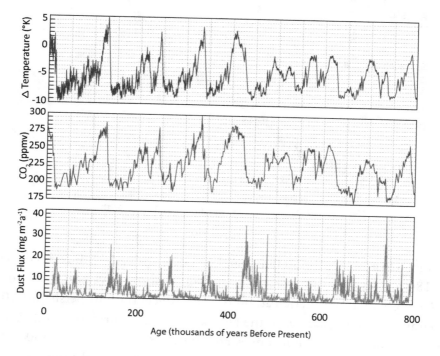

The regional sources of carbon dioxide emissions have changed over the last 20 years. (Table 3.5) North America, and especially the United States, remains a major source of

79 Oxygen 16 (O^{16}) is the predominant form of the oxygen atom on Earth. It has 8 neutrons and 8 protons. Oxygen 18 (O18) also exists in nature. It has 8 protons and 10 neutrons, so its atomic weight is greater.

80 https://commons.wikimedia.org/wiki/File:%22EDC_TempCO2Dust%22.svg

81 J. Shakun et al., "Global Warming Preceded by Increasing Carbon Dioxide Concentrations during the Last Deglaciation," *Nature*: 484 (2012): 49-54.

carbon emissions but has been surpassed by China. In the past 5 years, carbon dioxide emissions in the US have fallen, largely as a result of conversion from coal to natural gas for electrical power generation. In contrast, the industrialization of China has greatly increased its carbon dioxide emissions, and China is now the largest CO_2 emitter. India and the Middle East have also seen dramatic increases in CO_2 emissions over the last 40 years. Population and GDP growth in non-OECD countries will likely continue to increase global CO_2 emissions even if OECD countries reduce their emissions.[82]

Table 3.5: Carbon dioxide emissions in 1970 and 2017 by region. Estimates are ±10%.

Region	1970 Emissions (Gigatons CO_2)	2017 Emissions (Gigatons CO_2)	Percent Change (%) 1970-2017
US	4.8	5.8	21
China	0.85	10.8	1176
India	0.2	2.7	1250
Middle East	0.3	2.9	867
Japan	0.85	1.3	53
Africa	0.3	1.5	500
Europe	6.6	6.3	(5)
Latin America	0.25	1.5	600
OECD	10.6	13.9	31
Non-OECD	5.2	24.3	367
World	16.3	39.8	144

The effects of these carbon dioxide emissions on the global fast carbon cycle are shown in Table 3.6.

Other Greenhouse Gases

The concentrations of methane and nitrous oxide have also increased over the past 40 years while the levels of chlorofluorocarbons have begun to stabilize (Figure 3.20).[83,84] At the end of 2019, atmospheric methane concentrations were 1875 parts per billion (ppb),

---

82   The OECD is the Organization for Economic Co-operation and Development. It is an intergovern-mental organization founded in 1961 and has 37 member countries. All members are democracies with modern market economies.

83   M. Etminan et al., "Radiative Forcing of Carbon Dioxide, Methane and Nitrous Oxide: A Significant Revision of the Methane Radiative Forcing," *Geophys. Res. Lett.* 43 (2016): 12,614-12,623.

84   J. H. Butler and S. A. Montzka, "The NOAA Annual Greenhouse Gas Index (AGGI)," NOAA Global Monitoring Laboratory (2020), https://www.esrl.noaa.gov/gmd/aggi.html.

and nitrous oxide concentrations were 332 ppb.[85] The atmospheric CO_2 concentration at the end of 2019 was 412 ppm, and the concentration of CO_2-equivalents was 500 ppm.

Table 3.6: 2015 to 2017 fast carbon cycle sources and reservoirs.
Estimates are ±10%. Values are gigatonnes carbon/year.

Fast Carbon Cycle	2015	2016	2017
Carbon Source		Gigatonnes carbon /Year	
Fossil Fuel/Industry	9.68	9.74	9.87
Land Use Change	1.62	1.30	1.39
Total Carbon Source	11.30	11.04	11.26
Carbon Reservoir		Gigatonnes carbon /Year	
Atmosphere	6.19	6.04	4.64
Ocean	2.58	2.64	2.51
Vegetation	1.84	2.58	3.78
Total Carbon Reservoir	10.61	11.26	10.93

Total Greenhouse Gas emissions

Electrical power generation, industrial processes, and transportation fuel use account for more than 50% of greenhouse gas emissions. Burning fossil fuels for these and other uses accounts for about 75% of greenhouse gas emissions. In 2018 the total amount of CO_2-equivalents emitted from burning fossil fuels was 49.6 gigatonnes (54.6 gigatons).

The average human produces about 2.3 pounds of carbon dioxide from respiration per day or 839.5 pounds or 0.38 tonnes per year. There are 7.8 billion humans, so annual CO_2 emissions from human respiration amount to only about 2.9 gigatonnes or about 6% of annual CO_2-equivalent emissions from burning fossil fuels. Human methane emissions are about 1 million tonnes and are inconsequential.

There are about 1.5 billion cows on Earth, each of which produces about 100 kg of methane and 4000 kg carbon dioxide per year, resulting in annual methane emissions of 150 million tonnes and carbon dioxide emissions of 6 billion tonnes. A tonne of methane is equivalent to 28 tonnes of CO_2 (100-year GWP), so the annual amount of CO_2-equivalent emissions from cows is about 8.2 gigatonnes or about 16.5% of annual CO_2-equivalent emissions from burning fossil fuels.

---

85   "Parts per billion" or "ppb" is a very small concentration. It is 0.0000001%. For example, 1 ppb is equivalent to about 3 seconds in 100 years.

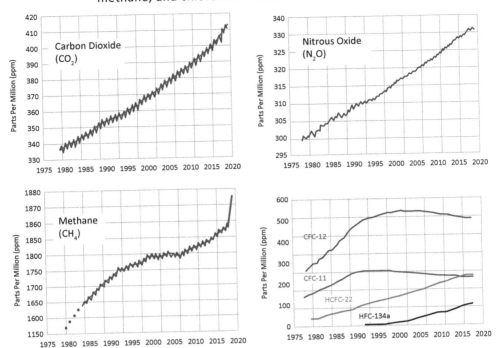

Figure 3.20: Concentrations of carbon dioxide, nitrous oxide, methane, and chloroflurocarbons from 1979 to 2018

The Earth's Energy Budget

If we take the daily total solar irradiance at the top of the atmosphere and distribute it uniformly across the surface of the Earth, the average daily solar irradiance is 340.4 watts per square meter of surface area per day (W/m²/day).[86] (Figure 3.21) Clouds and the atmosphere reflect 77.0 W/m²/day of solar irradiance back into space. The atmosphere absorbs 77.1 W/m²/day, allowing 186.2 W/m²/day to reach the surface. The surface reflects 22.9 W/m²/day, leaving 163.3 W/m²/day (48.4%) that is absorbed by the surface. About 92% of this energy is stored in the oceans because the oceans cover 71% of the Earth's surface, they are deep, and they have a high heat-absorption capacity compared to the atmosphere and land.

The Earth's surface, including the oceans, emits 398.2 W/m²/day, mostly in the infrared spectrum as heat. The atmosphere absorbs 358.2 W/m²/day of this terrestrial radiation and 40.1 W/m²/day passes through the atmosphere into space. In addition, evaporation of water from the Earth's surface, transpiration of water from plants, and thermal convection adds 104.8 W/m²/day to the atmosphere.

In summary, the atmosphere absorbs 77.1 W/m²/day from the sun, 358.2 W/m²/day

---

86   M. Wild, D. Folini, C. Schar, N. Loeb, E. G. Dutton, and G. Konig-Langlo, "The Global Energy Balance from a Surface Perspective," *Climate Dynamics* 40 (2013): 3107–3134, DOI 10.1007/s00382-012-1569-8.

from the surface, and 104.8 W/m²/day from convection and evapotranspiration. The total is 540.1 W/m²day. The atmosphere and clouds emit 199.8 W/m²/day into space and reradiate 340.3 W/m²/day back to the surface for a total of 540.1 W/m²/day.

The total amount of energy emitted by the surface is 398.2 W/m²/day plus 104.8 W/m²/day of convection and evapotranspiration for a total of 503.0 W/m²/day. The total amount of energy absorbed is 503.6 W/m²/day. Thus, the net amount of energy retained by the Earth is 503.6 W/m²/day absorbed − 503.0 W/m²/day emitted = 0.6 W/m²/day net heat retention, which is only 0.2% of global insolation. More recent measurement of these fluxes results in an estimated net absorption of 0.7 ± 0.5 W/m²/day.

ARGO measurements can be used to evaluate the net energy absorbed by Earth as well. In 2016, the ARGO program estimated the average net absorbed surface energy at 0.71 ± 0.10 W/m²/day. The data indicated that 0.61 W/m²/day was stored in the upper ocean (a depth up to 1,800 m), 0.07 W/m²/day was stored in the deep ocean (depth greater than 1,800 m), and 0.03 W/m²/day melted ice or was stored in the land or atmosphere. Importantly, the CERES and ARGO estimates were very similar, suggesting the average net energy absorbed by the Earth is likely to have been between 0.6 and 0.8 ± 0.5 watts/m²/day in recent years. We can convert the estimates of radiative forcing to joules per year. A watt is one joule per second, and the Earth's surface area is 510.1 million Km². This means that the Earth is now accumulating excess energy at the rate of about 10-15 zettajoules (10^{21} joules) per year. For comparison, in 2018, total worldwide human energy consumption was only about 0.6 ZJ. Accumulation of 10 zettajoules (ZJ) of energy each year is equivalent to accumulation of the energy in 1.6 trillion barrels of oil or nearly all of the world's proven oil reserves in 2019.

The Forecast:

During the remainder of the 21st century, Earth's energy balance will continue to change, as will the carbon cycle. These dynamics, which will drive climate change, are summarized in Table 3.7. Based on our current understanding, we can expect the level of solar radiation to remain stable through 2100. The Earth's albedo may decrease due to the loss of surface ice and snow, especially in the northern hemisphere, offset by an increase in albedo from more dense clouds in northern latitudes.[87] Volcanic activity could add aerosols and particulates to the stratosphere that would increase the Earth's albedo, but these changes would be short lived. The concentration of greenhouse gases is going to increase driven by population and GDP growth and the availability of low-cost fossil fuels in developing countries. The net effect of all of these changes will be an increase in radiative forcing. The absorption of energy by the Earth's oceans will increase and ocean temperatures will rise.

---

87   J. Norris et al., "Evidence for Climate Change in the Satellite Cloud Record," *Nature* 536 (2016): 72-75.

Changes in the Earth's carbon cycle are also likely. Human carbon emissions will likely increase over the next 50 years. Increased atmospheric carbon dioxide concentrations will increase photosynthetic activity but this effect will likely be reduced by desertification and deforestation. Ocean carbon dioxide uptake will increase which will cause ocean pH to decrease and could cause ocean photosynthetic activity to increase. If the tundra thaws due to global warming and stored methane is released into the atmosphere, the increase in the greenhouse effect by mid-century could be greater than expected due to the higher net radiative forcing effect of methane.

Since 1970 carbon dioxide concentrations in the atmosphere have gone up on average 1.8 ppm/year, and the global mean surface temperature anomaly has risen on average 0.018°C/year. This simple relationship (i.e., 0.01°C rise in global mean surface temperature per 1 ppm rise in atmospheric CO_2 concentration) would suggest that an increase in carbon dioxide levels from 291 ppm in 1900 to 500 ppm in 2050 would be associated with a 2.1°C increase in the global surface temperature anomaly during this same period. We have already experienced a 1.0°C increase since 1900, so we might expect an additional 1.1°C increase through the middle of the 21st century. In reality, the relationship between changes in atmospheric carbon dioxide concentrations and global surface temperatures is much more complex because of feedback in the climate system.

Figure 3.21: The Earth's energy budget

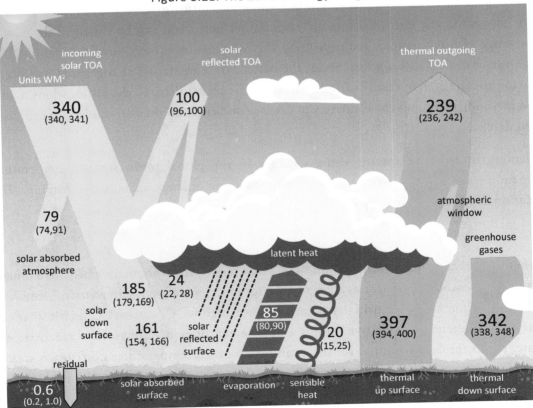

Table 3.7. Expected changes in the Earth's energy budget

Factor	Moderating Effects	Expected Change to 2100
Global Insolation	Solar Cycles, Milankovitch Effect	No major change
Earth's Albedo	Clouds, Aerosols, Particulates, Volcanic Activity, Surface Ice & Snow	No change or possible small decrease in Albedo due to: • Increased clouds • Decreased Surface Ice • Unchanged volcanic activity
Surface absorption of solar energy	Ocean Effect, Vegetation Effect	Increase
Greenhouse effect	Atmospheric Carbon dioxide, Ozone, Methane, Nitrous Oxide, and Chloroflurocarbon concentrations.	Large Increase due to: • Increase in clouds and water vapor • Increase in $[CO_2]$ • Increase in $[CH_4]$ • Increase in $[NO_2]$

An enormous effort has been made by international climate scientists to develop, test, and refine various mathematical models designed to provide forecasts of future changes in the global mean surface temperature. Assumptions about future greenhouse gas emissions and any changes in the Earth's albedo are essential elements and can be varied to determine the range of future surface temperatures. The temperature projections for 2050 or 2100 also vary based on assumptions about the pattern of temperature changes that will occur during climate system equilibration. These models are sensitive to the assumptions made, the methods used, and the estimates of future conditions so they are subject to criticism and debate.

Changes in radiative forcing cause changes in the global mean surface temperature depending on the sensitivity of the climate to these changes in radiative forcing. The relationship is relatively simple:

$$\Delta T_s = \lambda_s \Delta F.$$

Where ΔT_s is the change in global mean surface temperature anomaly when the climate system has reached equilibrium, ΔF is the change in radiative forcing in watts/m^2, and λ_s is the equilibrium climate sensitivity parameter measured in °C/watt/m^2. Climatic changes due to a difference in radiative forcing can take centuries or millennia to reach steady state. For example, changes in ice sheets and deep ocean temperatures don't reach steady state for many hundreds of years or millennia. Climate sensitivity is usually expressed as the change in the global mean surface temperature anomaly at equilibrium in response to a doubling of the CO_2 concentration in the atmosphere. Estimates of equilibrium climate sensitivity come from multiple studies of paleoclimate data, studies of major volcanic eruptions, studies of the effect of solar cycles on climate, and validation studies of several climate models over the past 30 or more years. The current estimate for equilibrium climate sensitivity to a doubling of atmospheric CO_2 concentrations is 3.0°C (Range: 2.6 to 3.9°C), although some estimates are higher.[88] Thus, the climate sensitivity parameter λ_s is estimated to be 0.81°C/watt/m² (Range: 0.70-1.05°C/watt/m²).[89] (See box below)

ΔT_s=Change in global mean surface air temperature anomaly at equilibrium (°C)

λ_s=Climate sensitivity parameter (°C/watt/m²)

ΔF=Change in radiative forcing due to changes in greenhouse gas concentrations (watt/m²)

α=Climate constant (watt/m²)

C=CO_2 concentration at time of interest (ppm)

C_0=CO_2 concentration at baseline or reference time (ppm)

$\Delta T_s = \lambda_s \Delta F$

$\Delta F = \alpha \ln(C/C_0)$

If α = 5.35watt/m²; and

ΔT_s = 3°C for C/C_0=2 then:

$3°C = \lambda_s [5.35 \text{ watt/m}^2 (\ln 2)]$

$\lambda_s = 0.809°C/watt/m^2$

Some changes in surface temperature occur within the first few decades after an increase in radiative forcing and are best estimated using a climate sensitivity measure termed the Transient Climate Response (TCR). TCR considers the change in global mean surface temperature that would occur if carbon dioxide levels increase 1% per year until they double. TCR is simply the global temperature increase that occurs at the time that atmospheric carbon dioxide levels double, keeping in mind that temperatures would

---

88   J. E. Tierney, J. Zhu, J. King, S. B. Malevich, G. J. Hakim, and C. J. Poulsen, "Glacial Cooling and Climate Sensitivity Revisited," *Nature* 584 (2020): 569-573, https://doi.org/10.1038/s41586-020-2617-x.

89   S. C. Sherwood, M. J. Webb, J. D. Annan, and K. C. Armour, "An Assessment of Earth's Climate Sensitivity Using Multiple Lines of Evidence," *Reviews of Geophysics* 58(4) (2020), https://doi.org/10.1029/2019RG000678.

continue to increase until equilibrium is reached. The current range of estimates for TCR based on historical data is 1.0°C to 2.6°C.[90]

The UN's Intergovernmental Panel on Climate Change (IPCC) has been the epicenter of discussion and debate about these future changes and their consequences. In order to facilitate and simplify the debate, the IPCC has developed a standardized approach to projecting future greenhouse gas emissions. First they convert all greenhouse gas emissions into "CO_2-equivalents."[91] In 2013, the IPCC created Representative Greenhouse Gas Concentration Pathways (RCPs) to represent different projected levels of radiative forcing through 2100. (Figure 3.22) For example, RCP 8.5 is the projected change in the concentration of greenhouse gases in the atmosphere that results in radiative forcing of 8.5 watts/m² in 2100. At the end of 2019, the atmospheric concentration of CO2-equivalents was 500 ppm, which is well above the RCP 8.5 projections.

Figure 3.22. Representative Greenhouse Gas Concentration Pathways (RCPs)

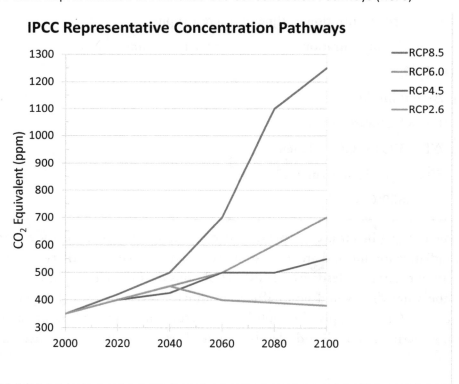

90   A. Schurer, G. Hegerl, A. Ribes, D. Polson, C. Morice, and S. Tett, "Estimating the Transient Climate Response from Observed Warming," *J Climate* 31 (2018): 8645-8663.

91   The CO_2-equivalent concentration is an estimate of the concentration of CO_2 that would be required to produce the same degree of radiative forcing as the sum of the effects of all greenhouse gases in the atmosphere.

By 2050, the world will likely be using about 0.75 to 0.85 ZJ of energy. If we assume that by 2050 fossil fuels will supply about 75% of this energy (as compared with 85% in 2019) and that coal will supply 20%, oil 25%, and natural gas 30%, then global carbon dioxide emissions from fossil fuels will be about 40-45 gigatonnes. If this output represents the same percent of total greenhouse gas emissions measured in CO_2-equivalents as it does today (about 66-75%), then we can expect total greenhouse gas emissions in 2050 to be 55-68 gigatonnes of carbon dioxide equivalents. This represents a 20% to 30% increase from 2019 emissions and would likely result in a pattern of change similar to RCP 8.5. RCP 6.0 projects CO_2-equivalent concentrations in 2050 of 500 ppm and in 2100 of 740 ppm and assumes the effect of other greenhouse gases is offset by the cooling effect of sulfate aerosols and clouds. This scenario seems highly unlikely given that the atmospheric CO_2-equivalent concentration at the end of 2019 was 500 ppm. RCP 8.5 predicts CO_2-equivalent levels of 640 and 1240 ppm, respectively, and may be the more likely prediction unless current trends are reversed.

Climate models predict that an increase in atmospheric carbon dioxide equivalent concentrations to 480 to 650 ppm by 2046-65 will likely result in a rise in global mean surface temperatures of 1.8°C to 3.2°C, relative to pre-industrial temperatures (13.74°C), with land warming about twice as much as oceans, and the northern hemisphere warming more than the southern hemisphere. (Table 3.8) These same models predict a rise in global mean surface temperatures in 2080-2100 of 2.4°C to 5.2°C, depending on the degree of climate sensitivity, with further warming expected in the 22nd century.[92]

Table 3.8 Predicted change in mean global surface temperature anomaly for RCP 4.5, 6.0, and 8.5

RCP Emissions Scenario	Atmospheric CO_2-equivalent Concentration (PPM)		Mean Temperature Change (± 1 Standard Deviation)	
	2046 to 2065	2081 to 2100	2046 to 2065	2081 to 2100
4.5	550	570	2.3 ± 0.3	2.8 ± 0.4
6.0	480	740	2.1 ± 0.3	3.2 ± 0.4
8.5	650	1100	2.9 ± 0.3	4.8 ± 0.4

The pattern of global warming is especially important but more difficult to predict. The geographic distribution of λ_s indicates that some regions are more responsive to local and remote forcing changes than others. Based on our current understanding, it's likely the

92   P. Brown and K. Caldeira, "Greater Future Global Warming Inferred from Earth's Recent Energy Budget," *Nature*: 552 (2017): 45-50.

rise in global temperatures will not be evenly distributed across the Earth. Some regions, especially in the northern latitudes, will warm more than southern latitudes. Although surface temperatures over land will be greater, ocean temperatures will also rise about 0.6° to 1°C (1.1 to 1.8°F) by 2050.

The north polar icecap will continue to melt, and sea levels will continue to rise. The salinity of the oceans will decrease but their acidity will increase.[93,94,95] The decrease in salinity and increase in acidity will lead to extensive loss of coral reefs and other important aquatic habitats; and much more aquatic species decline or extinction than we have already witnessed. Sea levels have already risen on average by about 8 inches over the last 100 years. Ocean levels could rise 10-20 inches or more during the next 80 years. A rise of 30 inches or more would have a devastating impact on coastal cities. Indonesia, China, Japan, Northern Europe, and the East Coast of the US would be particularly hard hit.

On land climactic changes will be even more severe. Glaciers and permafrost will continue to melt, especially in the northern hemisphere where most of the permafrost is located. Deserts will expand in Africa, Asia, and southwestern North America. Northward migration of clouds will likely continue and cause changes in the distribution of rainfall across the Earth. The loss of glaciation will cause decreased flow in several major rivers that are an important source of freshwater for large parts of Africa and Asia. Most of these effects have already begun to occur. According to NOAA, from 1980 until 2018 the mass of mountain glaciers worldwide decreased by 21.7 meters of water, the equivalent of slicing 24 meters (79 feet) of ice off the top of the average glacier. Even more disturbing is the fact that the pace of glacier loss has accelerated from a loss of 9 inches per year in the 1980s, to 17 inches per year in the 1990s, to 26 inches per year in the 2000s, to 36 inches per year for 2010-2018.

Some of these effects will accelerate climate change. For example, rising global temperatures cause an increase in atmospheric water vapor concentrations. Water vapor is a potent greenhouse gas so increased atmospheric water vapor concentrations would accelerate global warming. An increase in cloud cover, on the other hand, would be expected to reflect more incoming solar radiation and cool the surface, unless the clouds migrate to northern latitudes where the Earth receives less solar radiation. Melting permafrost could release large stores of organic carbon dioxide and methane from the soil into the

93   T. Frederikse, F. Landerer, L. Caron, et al., "The Causes of Sea-Level Rise since 1900," *Nature* 584 (2020): 393–397, https://doi.org/10.1038/s41586-020-2591-3.

94   N. Wunderling, M. Willeit, J. F. Donges, R. Winkleman, "Global Warming Due to Loss of Large Ice Masses and Arctic Summer Sea Ice," *Nature Comm.* 11 (2020): 5177, https://doi.org/10.1038/s41467-020-18934-3.

95   N. Mimura, "Sea-Level Rise Caused by Climate Change and Its Implications for Society," *Proc. Jpn. Acad. Ser. B Phys. Biol. Sci.* 25; 89(7) (2013): 281–301, doi: 10.2183/pjab.89.281.

atmosphere, adding to the greenhouse effect. Warming and acidification of the oceans reduces the capacity of oceans to store excess atmospheric carbon dioxide and increases the greenhouse effect. Finally, as the amount of arable land in lower latitudes decreases, deforestation to expand the availability of land for agriculture could reduce the amount of carbon dioxide removed from the atmosphere by photosynthesis. Some of these interactive effects could be significant.

If climactic zones shift, sub-arctic zones may become suitable for agriculture, including large areas of Alaska, Canada, and Russia, although solar irradiance will be much more variable during the year at higher latitudes. Plants need sunlight, water, nutrients, and carbon dioxide to grow. Each plant species, including agricultural crops such as wheat, corn, rice, and soybeans, has an optimum growing range. Crop yields usually decrease as conditions deviate above or below the optimum range. As atmospheric carbon dioxide concentrations rise, agricultural yields would be expected to increase because there is more carbon dioxide available for the light reactions of photosynthesis. If clouds increase as the climate warms, the amount of photosynthetically active radiation (PAR) could decrease, depending on where the clouds are. If PAR decreases, crop yields would be expected to decrease because light is needed for plant growth. If agricultural zones shift to higher latitudes, annual crop yields will be reduced because of light availability, especially if the amount of shading from clouds increases. Rising temperatures may also have an effect on crop yields, especially if the optimum growing range is exceeded for much of the day in existing agricultural zones.

Modeling photosynthesis and predicting crop yields can be a challenge, even under controlled conditions. Doing so on a global basis, particularly when there may be significant uncertainty about the input variables, is especially difficult. Input variables include atmospheric carbon dioxide concentrations, PAR, water availability, temperatures during the day and night, and nutrient availability (phosphate, nitrate, calcium, etc.). In order to keep the models manageable, some growing conditions are usually assumed to be constant (e.g., availability of nutrients and water). Each model is crop specific. Unfortunately, current climate models do not perform well in predicting some of these input variables, such as rainfall and cloud cover. Finally, some input variables in these models may interact and estimating the magnitude of the interaction may be problematic, especially if several variables are included in the model. As a result, at present, results from agricultural productivity models should only be regarded as indicators of the potential direction of future crop yields.

If carbon dioxide concentrations increase to about 500 ppm by 2050 and the rise in global surface temperatures is below 2° C, then crop yields in most agricultural zones are likely to increase slightly. If more extreme conditions occur and global surface temperatures rise by 3-5° C, resulting in prolonged droughts or floods in existing agricultural regions, or sustained daily temperatures above the optimum growing range, crop yields could be reduced. There is currently enough arable land, water, and fertilizer available to grow enough food to meet the nutritional demands of the world's population, yet about 800 million people go hungry, especially the rural poor in under-developed countries. The main reasons for this are regional weather conditions (drought, extreme storms), war and political conflict, a lack of modern agricultural technology (fertilizer, equipment, herbicides, pesticides, high yield seed, etc.), and education in proper farming techniques. The demand for food is expected to double by the end of the century due to population growth, especially in Africa, India, and the Middle East. Unfortunately, these same regions are expected to experience some of the largest increases in water stress through the end of the century. Couple the lack of water with the lack of modern agricultural technology, lack of education, political conflict, and the potential for extreme weather conditions, and the prospects for these regions being able to sustain the population growth they are likely to experience look grim. Should this be the case, mass migration from these regions may ensue.

A Brief History

The insight that burning fossil fuels could cause the global mean surface temperature to increase was first published by the Swedish physical chemist Svante Arrhenius in 1896. In his paper "On the Influence of Carbonic Acid in the Air on the Temperature of the Ground," he states that the atmosphere absorbs heat chiefly through water vapor and carbonic acid (H_2CO_3 or aqueous CO_2).[96] He proposed an equation that relates carbonic acid concentrations to the change in surface temperature. This equation is still used today. The modern formulation of Arrhenius' equation is:

$$\Delta F = \alpha \ln(C/C_0)$$

ΔF is the change in radiative forcing, α is a constant estimated from observation, C is the current concentration of CO_2 in the atmosphere and C_0 is the CO_2 concentration at some prior reference point in time, usually pre-industrial levels of CO_2. The current best estimate for α is 5.27 to 5.35 watts/m^2, depending on atmospheric carbon dioxide and nitrous oxide concentrations. There is some uncertainty about this estimate, but the level of uncertainty is low (less than 10%).

Arrhenius concluded that a 2.5- to 3-fold increase in atmospheric carbon dioxide concentrations would result in an increase in arctic temperatures of 8° to 9°C. While he discussed the effect of burning coal on carbonic acid concentrations, his main conclusion related to explaining geologic climatology and the conditions that produced ice ages. At

---

96   Svante Arrhenius, "On the Influence of Carbonic Acid in the Air upon the Temperature of the Ground," *London, Edinburgh, and Dublin Philosophical Magazine and Journal of Science* (fifth series) 41 (1896): 237–275.

the time, ice ages were thought to be due to variations in the Earth's orbit around the sun. In a later paper he predicted that emissions of carbon dioxide from burning fossil fuels would be sufficient to cause global warming and predicted that doubling the concentration of CO_2 in the atmosphere would cause a rise in global mean surface temperatures of 4°C.

Guy Callendar was an English engineer and amateur climatologist who read the work of Arrhenius and other early 20th century climatologists and began compiling climatological data in the 1930s. In 1938 he published "The Artificial Production of Carbon Dioxide and Its Influence on Temperature" in the *Quarterly Journal of the Royal Meteorological Society*.[97] In this paper, Callendar related changes in atmospheric carbon dioxide concentrations to changes in surface temperatures over the previous 50 years. He observed that mankind had added 150 megatons of carbon dioxide to the atmosphere during this period and that global land temperatures had increased on average 0.005°C/year. He proposed that this increase in temperature could be explained as an effect of the increase in atmospheric carbon dioxide concentrations. Callendar proposed that the value for climate sensitivity was 2°C, a level that is within the range of current estimates.

Charles Keeling was an American chemist who worked at the Scripps Institute of Oceanography. As a graduate student at Caltech, he developed an instrument for accurately measuring carbon dioxide in water and air. After conducting exploratory measurements at ground level, he recognized the influence of nocturnal plant respiration and temperature changes on his measurements. To minimize or eliminate these influences, in 1958 he established an observatory on the peak of the Mauna Loa volcano in Hawaii to take direct measurements of atmospheric CO_2 concentrations. In 1968 he reported his first six years of data in the *Journal of Geophysical Research*.[98] He concluded "The CO_2 concentration at Mauna Loa Observatory varies with season with an average amplitude of 6 ppm and is increasing at the average rate of 0.7 ppm per year. Volcanic emanations of CO_2 near the summit of Mauna Loa and uptake of CO_2 on the forested lower slopes of the mountain influence the concentration of CO_2 at Mauna Loa Observatory but do not seriously interfere with the determination of regional changes." Keeling showed in what later became known as a "Keeling Curve" that the concentration of carbon dioxide in the atmosphere was steadily increasing. President Johnson's Science Advisory Committee used Keeling's data to warn of the potential future dangers of greenhouse gas emissions.

In the 1970s, based on Keeling's observations and the accumulating evidence for climate change, the international scientific community began to receive financial support for

---

97   G. S. Callendar, "The Artificial Production of Carbon Dioxide and Its Influence on Temperature," *Quarterly Journal of the Royal Meteorological Society* (1938), doi: 10.1002/qj.49706427503.

98   C. D. Keeling et al., "Concentration of Atmospheric Carbon Dioxide at 500 and 700 Millibars," *J. Geophys. Res.* 73 (1968): 4511–4528.

expanding and improving the gathering of climate data. The number of weather stations increased; orbiting satellites, ocean buoys, and weather gathering ships were deployed; and cooperative multi-national research programs were organized to gather, analyze, and report the results. James Hansen, working for the Goddard Institute for Space Studies (GISS), began working on the influence of aerosols and gases on Venus's and Earth's climate. Equipped with richer and more accurate data sets and improving computer technology, Hansen and his coworkers published a climate model in 1981, "Climatic Impact of Increasing Atmospheric Carbon Dioxide" in *Science*.[99] The results indicated that meaningful global temperature data could be gathered from land-based stations and that an increase in global surface temperatures of about 0.5° to 0.7°C had occurred in the past century, with warming of similar magnitude in both hemispheres.

During the 1980s, work continued on the methodology for analyzing satellite and weather station data. Paleoclimate data (e.g., ice core data) began to emerge that validated Milankovitch's calculations but showed that the changes in surface temperatures could not be explained simply by changes in orbital forcing. In 1988 The Intergovernmental Panel on Climate Change (IPCC) was established by the United Nations and the World Meteorological Organization (WMO) to provide expert opinions, data analysis, and policy recommendations to mitigate future climate changes.

By the 1990s, computers had become much more powerful and could now easily handle the complex calculations needed for climate modeling. The number and complexity of the climate models grew. Vigorous debate, differences of opinion, and skepticism about the results of the models ensued. In December 1997, an international protocol to address climate change was adopted in Kyoto, Japan. The Kyoto protocol set internationally binding emission targets that were heavily weighted toward the OECD countries, recognizing that most of the prior carbon emissions had been generated by the developed world. The protocol was based on the principle of "common but differentiated responsibilities." Detailed rules for implementation were adopted in 2001 in Marrakesh (the Marrakesh Accords), and the first commitment period started in 2008 and ended in 2012. Thirty-eight countries agreed to participate, although the two largest emitters, the US and China, did not. By 2012 most participating countries had not met their greenhouse gas emissions targets. However, the Kyoto protocol was a first step toward a coordinated international effort to reduce greenhouse gas emissions.

Throughout the first decade of the 21st century, scientists continued to collect data, revise prior estimates of the effect of greenhouse gases on radiative forcing and validate prior estimates of climate sensitivity. There were many published validation studies of climate models and most performed well using a variety of statistical approaches to

---

99   J. Hansen, D. Johnson, A. Lacis, S. Lebedeff, P. Lee, D. Rind, and G. Russell, "Climate Impact of Increasing Atmospheric Carbon Dioxide," *Science*, 213 (1981): 957-966.

validation. Many of the published scientific results actually increased the expected impact of greenhouse gas emissions and estimated changes in global mean surface temperatures through mid-century or 2100. The environmental movement embraced the projections and began efforts to popularize global warming and raise public awareness of the threats. In 2006 a documentary film called "An Inconvenient Truth" was released and received a great deal of public attention since the former Vice President of the United States, Al Gore, was a main participant. This documentary raised public awareness of the threat of global warming and other potential climate changes. It also served to galvanize political views regarding climate change projections. Several powerful interests who saw the science and climatic projections as a threat made concerted efforts to discredit the science and scientists and/or undermine efforts to develop effective mitigation strategies. These efforts are ongoing.

The most recent international effort to develop and implement effective strategies to lower greenhouse gas emissions is the 2015 Paris Agreement.[100] This agreement builds on the principles established in Kyoto but seeks to involve all nations in a coordinated effort to limit global warming to 2°C or less by the end of the 21st century. The agreement also aims to strengthen the ability of all countries to adapt to the effects of climate change. The Paris Agreement requires all Parties to put forward their best efforts through nationally determined contributions (NDCs). This includes requirements that all Parties report regularly on their emissions and on their implementation efforts. By 2015, 195 countries had signed the Accord. However, in 2018 the US withdrew from the agreement. As of the end of 2019, nearly all signatories were falling short of their goals. Of the 32 signatories that account for ~80% of the world's GHG emissions, only 7 had made commitments or efforts that would meet the Paris targets and half were either "critically" or "highly insufficient" in meeting their goals, including the US and China.

Conclusion:

Over the last century mankind has learned a great deal about the Earth's energy balance, carbon cycle, and climate, and how they relate to each other. Scientists have been able to develop a good understanding of how the climate system works, how the different components fit together, and how the system responds to changes. Of course, there is still vigorous debate about the scientific details, a need for a more precise understanding of the quantitative relationships, and a lot to learn. Nevertheless, the body of work has led us to the realization that mankind has perturbed the Earth's energy balance and carbon cycle by burning fossil fuels and releasing greenhouse gases into the atmosphere. As a result, the amount of solar radiation retained on the Earth's surface has increased, especially the amount retained in the oceans. These changes are not evenly distributed; some regions of

---

100   The Paris Agreement, https://Unfccc.int.

the world are much more severely affected than others. The magnitude and duration of these effects suggest that by the end of this century the climatic response is likely to pose a serious threat to many habitats, including our own.

Summary:

The Earth receives the vast majority of its energy from the sun. The amount of energy absorbed by the Earth is a function of four major factors: the amount of solar radiation reaching the top of the Earth's atmosphere, the Earth's albedo, the heat capacity of the Earth's surface, and the effect of the Earth's atmosphere, including the greenhouse effect. Radiative forcing is the net amount of solar energy retained by the Earth each day. Radiative forcing is currently estimated to be 1.5 ± 0.8 watts/m^2 of Earth's surface. The oceans absorb most of this energy. As radiative forcing increases, the greater heat absorption capacity of the oceans moderates the increase in global mean surface temperatures. If radiative forcing decreases, heat transfer from the oceans will reduce the rate of global cooling. Thus the Earth's oceans serve as a major determining factor of global surface temperatures.

The greenhouse effect is caused primarily by water vapor, carbon dioxide, methane, and nitrous oxide in the atmosphere. These gasses absorb energy from the sun and from the Earth's surface and reradiate it as infrared energy (or heat) back toward the Earth's surface. An increase in the concentration of greenhouse gases in the atmosphere leads to an increase in radiative forcing. The climate sensitivity of the Earth to a doubling of the atmospheric carbon dioxide concentration is currently estimated to be 3.0°C at equilibrium (Range: 2.6 to 3.9°C) but could be higher.

The average surface temperature of the Earth has increased by 1.0°C $\pm$ 0.05°C over the last 120 years. This temperature increase is primarily due to an increase in the greenhouse effect. It cannot be explained by changes in solar cycles or changes in the Earth's albedo. Most of the increase in the greenhouse effect is due to carbon dioxide emissions from burning fossil fuels. Over the next 80 years no significant change is expected in solar radiation cycles or in the amount of solar radiation reaching the Earth. The greenhouse effect is expected to increase, which will change the Earth's energy balance by increasing the level of radiative forcing. By 2050 the atmospheric concentration of CO_2-equivalents is likely to exceed 550 ppm unless carbon mitigation strategies substantially lower carbon emissions. By the end of the 21st century, the expected changes in the Earth's energy budget will likely result in an increase in the global mean surface temperature of 3.8°C $\pm$ 1.4°C, relative to pre-industrial levels, with further warming expected in the 22nd century.

The change in surface temperatures will not be evenly distributed. Land will warm at least twice as much as the oceans and the northern hemisphere will warm more than the southern hemisphere. High northern latitudes are likely to experience the largest increase in temperature. How the oceans respond to the rise in temperature will determine many of the climatic consequences, especially if rising ocean temperatures alter major ocean currents. Glaciers and both polar ice caps will continue to melt, raising sea levels, lowering the salinity of oceans and decreasing water flow in many major rivers of the world. The amount of water vapor in the atmosphere and the amount of cloud cover will increase, especially at the equator and northern latitudes. Cloud and precipitation patterns will change, which will likely increase desertification, especially in northern Africa and Central Asia. Precipitation in high northern and southern latitudes will increase. Climate change and global warming may alter the El Niño Southern Oscillation. Major storms will increase in intensity driven by warmer ocean temperatures. Some regions of the world will experience dramatic changes in their climate that will alter the availability of fresh water and the amount of arable land. The social, political, and economic consequences of global warming are likely to be profound.

CHAPTER 4

Natural Resource Depletion

*"Since natural resources are finite, increased consumption
must inevitably lead to depletion and scarcity."*
—Paul Ehrlich

Chapter Guide

THIS CHAPTER DESCRIBES THE DYNAMIC relationship between the world's current reserves of essential natural resources and future demand for these resources, including fossil fuels, fertile land, and freshwater. After reading this chapter you should understand the relationship between population/economic growth and resource demand, including the relationship between current estimated total terrestrial fossil fuel reserves and future energy demand. It is especially important to understand how the reserves of essential natural resources differ from country to country and region to region and how climate change and population growth will cause scarcity in many regions. Understanding the magnitude of the need and the cost associated with replacing non-renewable with renewable energy resources and understanding the challenges associated with replacing losses of arable land and groundwater are especially important. A summary concludes the chapter.

Introduction

The natural world provides a wide array of essential resources, such as sunlight, minerals, water, biomass, and land. Some of these resources are perpetual, such as solar and wind energy; some are renewable, such as biomass and fresh water; and some are finite and consumption can lead to depletion, such as fossil fuels and minerals. Estimating the future availability of finite natural resources is a complex undertaking. The resource-related factors to be considered include the magnitude of current reserves, the amount of any non-renewable resource previously extracted, current and future demand, availability of alternative resources or new technology, and the economics of extracting and distributing the resource. Knowledge of these factors may be used to estimate the time frame and pattern of depletion, at least over the long-term.

Energy

The majority of the energy currently used by mankind comes from burning fossil fuels: oil, natural gas, and coal. These fuels are ancient biomass that has been converted into hydrocarbons by the temperature and pressure of the Earth's mantle. While the Earth's stores of these fuels are very large, they are finite, and they are not evenly distributed geographically. For millennia humans have used natural resources to generate energy and do work. The resources used have included wood, charcoal, straw, dung, peat, animal or vegetable oils, wind, and water. Biomass in one form or another was also used as feed for livestock that either did work or served as food. Fire provided heat and light but the energy it produced couldn't be stored or easily transported. Simple machines, driven by wind or water, ground grain and pumped water. Animals and sailing vessels were the primary modes of transportation.

The invention of the steam engine in 1698 greatly expanded the use of coal in the 18th century and ushered in the industrial revolution. The invention of the electrical generator by Faraday in 1831 accelerated industrial development. The development of the internal combustion engine in the mid-nineteenth century led to widespread use of oil, which was refined into gasoline. By the dawn of the 20th century, cheap, readily available energy derived from fossil fuels made modern agriculture, cities, and commerce possible. In 1954 the first nuclear power plant was connected to the electrical grid in Russia and nuclear power was added to the list of major energy sources.

An ideal energy source is one that is inexpensive, safe to use, reliable, and has a positive net energy balance. The net energy balance is the amount of energy produced minus the amount of energy needed to produce it. Because mankind now uses enormous amounts of energy, an ideal energy source must be scalable to very high production levels. Modern transportation needs require that an energy source be transportable or at least storable in a form that can be transported. Cost considerations include the cost of maintenance and the cost of "down time" if the energy source isn't reliable. Given the scale of current energy demand, efficiency is a critical factor. Finally, an ideal energy source should have a low environmental impact, require little or no other natural resources for production, and should be renewable or perpetual. All of the currently available energy sources have advantages and disadvantages. Fossil fuels (oil, natural gas, and coal) meet most of the criteria of an ideal energy source. They are cheap, storable, transportable, safe, reliable, efficient, scalable, low maintenance and have a high return on the energy invested to produce the fuel. Unfortunately they are not renewable and burning them to produce energy can have serious adverse environmental effects.

Renewable energy sources can be used to generate electricity, including hydroelectric plants, solar voltaic and solar thermal power plants, wind turbines, and geothermal power plants. Electricity is generally safe but requires a well-designed, secure, and expensive distribution system that can be inefficient if not well maintained. Solar and wind power generation can be unreliable if the amount of sunlight or wind falls below required thresholds. Batteries are needed to store electricity, especially electricity that is generated when production exceeds demand. In the absence of a high capacity and efficient storage capability, electricity can be an inefficient method of energy production. In order to use electricity for transportation, high capacity, reliable, durable and inexpensive batteries are needed.

Nuclear energy is also used to generate electricity. In addition to the considerations noted above, nuclear energy poses a unique problem; it must be fail-safe, or the adverse health and environmental consequences can be catastrophic and very long lasting. Hydrogen is an attractive energy source because it can be burned without producing carbon emissions. Hydrogen fuel cells were first invented in 1839. These fuel cells react hydrogen with oxygen in the presence of a catalyst to produce an electrical current and water. Fuel cells are efficient but are costly and can pose safety concerns because hydrogen is highly flammable.

Figure 4.1 depicts the consumption of these energy resources since the industrial revolution.[101] A terawatt-hour and an exajoule are units of energy (see Glossary). One hundred thousand terawatt-hours are equal to 360 exajoules, which is the amount of energy obtained by burning about 59 billion barrels of oil. Energy consumption greatly accelerated after WWII as a result of population growth and the availability of cheap fossil fuels. A very high proportion of the energy consumed after 1950 has been derived from fossil fuels.

World energy consumption is driven primarily by population growth and growth in world gross domestic product (GDP), offset by improvements in energy efficiency. World energy consumption by fuel type is shown in Table 4.1.[102]

In 2018, about 30% of global energy consumption was used for agriculture and about 20% for transportation; the remaining 50% was used for commercial, industrial, or residential purposes. The proportion of world energy consumption derived from fossil fuels has remained relatively constant at about 85% for the past several decades, despite vigorous efforts to develop renewable sources such a solar, wind, and biomass. Petroleum accounts for about 34% of global energy consumption, coal for about 27%, and natural gas for about 24%.[103] Nuclear power has not grown significantly in recent years, largely because of the environmental risks associated with nuclear power generation. Nuclear

101 H. Ritchie and M. Rosner, "Global direct primary energy consumption" (2020), https://ourworldin-data.org/grapher/global-primary-energy?time=earliest..latest.

102 British Petroleum Company, *BP Statistical Review of World Energy*, 69th edition (London: British Petroleum Co., 2020).

103 British Petroleum Company, *BP Statistical Review of World Energy*.

energy sources account for only about 4% of global usage. Renewables, such as biomass, wind, solar, and hydroelectric, account for about 11% of energy usage. Burning biomass accounts for most of the energy generated from renewable sources.

Figure 4.1: World energy consumption from 1800 to 2019 by fuel type

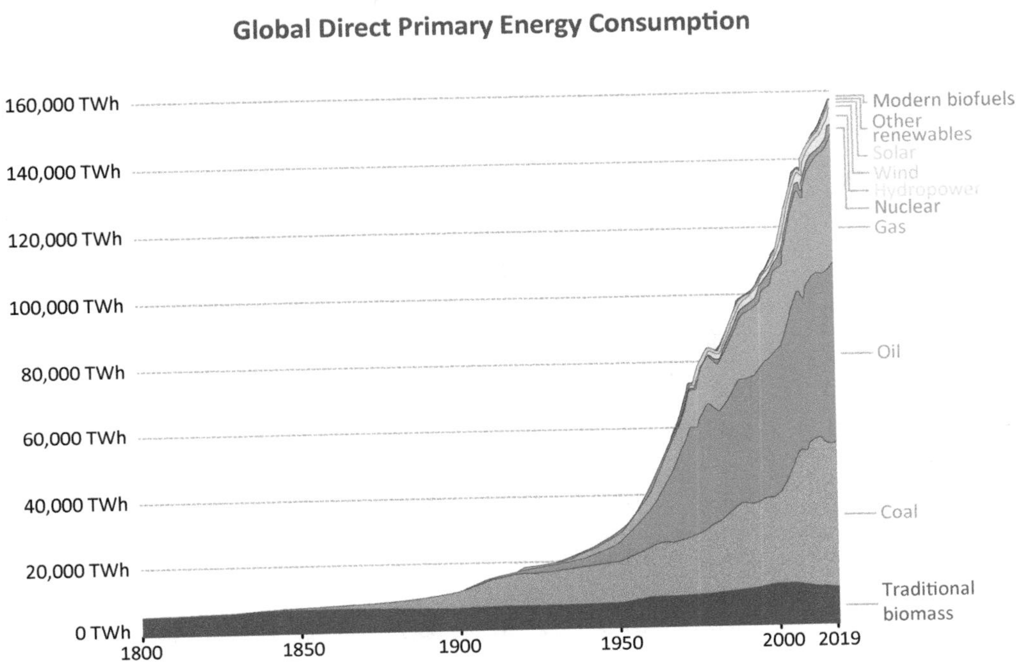

Global Direct Primary Energy Consumption

In 2019, US energy consumption was 105.7 EJ or 100.2 quadrillion British thermal units (Btu).[104] (Figure 4.2) The sectors driving consumption were as follows: electricity generation (37.8%), transportation (27.9%), industrial (22.7%), residential (6.8%), and commercial (4.6%). US domestic energy production in 2019 was 106.6 EJ. Fossil fuels accounted for 80% of primary US energy production in 2019; natural gas accounted for 32%, oil for 37%, and coal for just 11%. Renewables produced 11% and nuclear power plants generated 8%. More cost-effective drilling and production technologies for oil and natural gas, especially enhanced oil recovery and fracking, have helped to significantly increase US production in the past few years, especially in Texas and North Dakota.

---

104   US Energy Information Administration, *Monthly Energy Review* (2019), https://www.eia.gov/energy-explained/what-is-energy/sources-of-energy.php.

Table 4.1. 2018 World energy consumption by energy source

Region	Energy Source (EJ)						Total Energy** (EJ)	Increase 2017 to 2018 (%)
	Oil	Natural Gas	Coal	Nuclear	Hydro	Renewable		
North America*	46.6	36.8	14.4	9.1	6.7	5.0	118.6	2.8
Russia/CIS	8.1	20.9	5.6	2.0	2.3	0.02	39.0	4.4
China	27.8	10.3	80.1	2.8	11.4	6.0	138.4	4.3
Middle East	17.3	19.9	0.3	0.07	0.14	0.07	37.8	2.4
Europe	31.1	19.8	12.9	8.9	6.1	7.2	85.9	<0.05
Africa	8.0	5.4	4.2	0.1	1.3	0.3	19.3	2.9
Latin America	13.2	6.1	1.5	0.2	6.9	1.5	29.4	0.3
India	10.0	2.1	18.9	0.4	1.3	1.2	33.9	7.9
Total World (EJ)	195.2	138.6	157.9	25.6	39.7	23.5	580.5	2.9
Total World (%)	33.6	23.9	27.2	4.4	6.8	4.0	100	

*Includes Mexico, Canada, and the US
**Total may not equal the sum of the sources because of rounding.

Four countries, China (23.6%), the US (20.2%), India (5.8%) and Russia (5.2%), with about 42% of the world's population, accounted for about 55% of global energy consumption in 2018. Europe (14.8%) and Japan (3.3%) accounted for an additional 18%. From 2006 to 2016, energy consumption in the US, Europe, Japan, and Russia either declined or grew less than 0.03% per year, while in China and India consumption grew greater than 4.0% per year. The US, China and Europe account for over 50% of the world's oil consumption. The US uses almost twice as much natural gas as any other country or region and China dominates the coal market, accounting for more than half of annual global consumption.

Figure 4.2: US energy consumption by source, 2019

U.S. Energy Consumption by Source, 2019

biomass 5.0% *renewable* heating, electricity, transportation	**biopetroleum** 36.7% *nonrenewable* transportation, manufacturing, electricity
hydropower 2.5% *renewable* electricity	**natural gas** 32.0% *nonrenewable* heating, manufacturing, electricity, transportation
wind 2.7% *renewable* electricity	**coal** 11.3% *nonrenewable* electricity, manufacturing
solar 1.0% *renewable* heating, electricity	**nuclear** (from uranium) 8.4% *nonrenewable* electricity
geothermal 0.2% *renewable* heating, electricity	

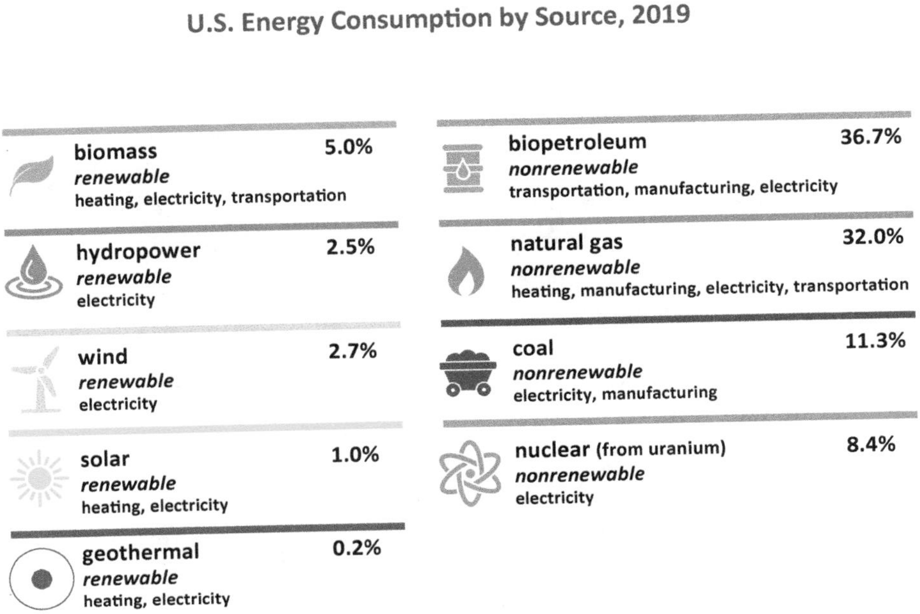

Future Energy Demand

From 1990 to 2017 world energy consumption grew on average 1.7% per year. However, world primary energy consumption growth jumped to 2.2% in 2017 and to 2.9% in 2018, up from 1.2% in 2016. In 2019 growth slowed to 1.3%. From 2006 to 2016, total energy consumption among the OECD countries decreased by 0.2% and in non-OECD countries grew by 3.3%. If we assume world energy consumption continues to grow at 1.7% per year, world energy consumption will be ~1.0 ZJ in 2050. If we assume improvement in energy efficiency reduces the growth rate to 1% per year, then 2050 consumption will be ~0.8 ZJ. To achieve a consistent and durable 1 % growth rate, global energy efficiency will need to improve by at least 40%, including in many developing countries that will experience significant population growth in the next 30 years.

The world population at the end of 2019 was about 7.7 ± 0.4 billion. World energy consumption in 2019 was 0.584 zettajoules (ZJ), which equates to global per capita energy consumption of 75.8 gigajoules (GJ = 10^9 joules) per person per year. If we project the world population to be 9.6 billion in 2050 and per capita energy consumption remains constant at 2019 levels, in 2050 world energy consumption will be 0.73 ZJ. In 2017 world per capita energy consumption grew by 2.2% but growth since 1970 has been about 1% per year. If we assume 1% growth per year, in 2050 per capita energy consumption will be 106.5 GJ per person. At this consumption rate, 2050 world energy consumption will be ~1.0 ZJ.

In 2016 global per capita GDP was $10,190 per person and global energy efficiency, defined as the amount of energy needed to generate a dollar of GDP, was 7.7 MJ/$GDP (megajoule or MJ = 10^6 joules). World GDP is expected to grow about 2.6% annually through 2050. At this growth rate, by 2050 world GDP will grow to ~$182 trillion, which equates to per capita GDP of ~$18,958. Energy efficiency has been improving for the past 30 years by ~2% per year. If this trend continues through 2050, global energy efficiency will be 4.1 MJ/$GDP. At this level of energy efficiency, world energy consumption in 2050 would be 0.78 ZJ. Unfortunately, improvements in energy efficiency in 2018 and 2019 were only about 1.2-1.5%, suggesting the rate of improvement in energy efficiency may be slowing. If so, world energy consumption in 2050 could exceed 0.85 ZJ.

Factors that are associated with an increase in per capita energy consumption are urbanization, migration to high-energy consuming countries, development of new energy intensive technology, and global warming.[105] Factors that cause per capita energy consumption to decrease are development of more energy efficient technology, a shift to less energy intensive economic activities, lifestyle changes or governmental policies that cause energy conservation or an increase in the cost of energy. Over the past few decades, growth in energy demand has closely followed growth in per capita income in low and middle-income economies, whereas high-income economies like the U.S. and Europe have produced GDP growth with little, if any, increase in per capita energy consumption. India, Brazil, Malaysia, and South Korea are good examples of low and middle-income economies where per capita GDP growth and per capita energy consumption have increased over the past 30 years. Thus, if the global population grows and per capita energy consumption increases, especially in India, Africa, and the Middle East, global energy consumption can be expected to increase.

The key to accurately projecting future energy demand in 2050 is estimating the rate of population and economic growth in non-OECD countries and the level of improvement in energy efficiency in these countries over the next 30 years. The US Energy Information Administration projects a modest 28% increase in world energy consumption by 2050, which equates to 0.76 ZJ. This predicted consumption rate essentially assumes the effect of growth in world population in the next 30 years will be offset by improvements in energy efficiency and that growth in world GDP will not significantly increase global energy consumption. A decrease in per capita energy consumption during a period when urbanization and global surface temperatures are expected to rise can only occur with improvements in energy efficiency that may be unrealistic.

Under most plausible assumptions about global population and economic growth and energy efficiency improvements, world energy consumption is likely to increase by about

---

105   B. J. van Ruijven, E. De Cian, and I. Sue Wing, "Amplification of Future Energy Demand Growth Due to Climate Change," *Nat Commun* 10 (2019): 2762, https://doi.org/10.1038/s41467-019-10399-3.

6 to 10 EJ per year. By 2050 world energy consumption is likely to be between 0.75 and 0.9 ZJ. If so, the world will need to cumulatively produce 20 to 22 ZJ of energy between now and 2050. By 2100 world energy consumption could exceed 1.1 ZJ per year. If world energy consumption grows at 1% per year, the world will need to generate ~80 ZJ of energy by the end of the century. If the climate warms more than 2°C by 2100, per capita energy consumption could increase by 25% or more and global energy consumption could be even greater.

Energy Production:

The energy content of fossil fuels varies. Coal contains 28-38 GJ/tonne, crude oil 42-46 GJ/tonne, and natural gas 52-56 GJ/tonne. For comparison, wood contains 18-22 GJ/tonne and charcoal 32-35 GJ/tonne. A thousand cubic feet (mcf) of natural gas contains ~1.1 GJ and a barrel of oil contains ~6.0 GJ.

"Proven" fossil fuel reserves are stores that have been previously located and can be recovered with a high probability (>90%) of success. "Recoverable" reserves have been located but have technical or other barriers to recovery that lowers the likelihood of success to an average of about 50%. "Undiscovered" reserves have been estimated from geological data. These reserves have a low probability of full recovery (less than10%) and include reserves that are unrecoverable with current technology. The current world proven reserves estimated by British Petroleum in its June 2020 Statistical Review of World Energy and converted into ZJ of energy are shown in Table 5.3.[106] In general, these estimates do not include landmasses that are sub-glacial.

"Unconventional" oil includes heavy oil, tight oil, tar sands, oil shale, and synthetic oil. Synthetic oil can be made from animal and plant oils, biomass, natural gas, and coal. Unconventional natural gas includes shale gas, coal bed methane, and tight gas. Tight oil and gas are oils or natural gas produced from reservoir rocks with such low permeability that substantial hydraulic fracturing (fracking) is necessary to produce them economically. Unconventional resources represent potentially very large reserves, although recovery is costly with current technology and has adverse environmental impacts. Estimates of world unconventional energy reserves may be imprecise or subject to considerable error. It is probable that much of these reserves, if they actually exist, would be very expensive or require new technology to fully exploit. Those unconventional sources that can be recovered with reasonable certainty and under current economic conditions are included in the proven reserves. Unconventional sources that are uncertain or would be unprofitable under current economic conditions are not included in proven reserves. The "total world reserves high estimate" in Table 4.2 includes the estimates of unconventional reserves that are not included in the proven reserves.

---

106 British Petroleum Company, *BP Statistical Review of World Energy.*

Table 4.2 Estimates of world conventional fossil fuel reserves. Estimates are from the *2019 BP Statistical Review of World Energy* and from the US Geological Survey estimate of world undiscovered fossil fuel reserves.

Fuel Source	2019 World Proven Reserves (ZJ)	Total World Reserves Low Estimate (ZJ)*	Total World Reserves High Estimate (ZJ)**
Oil	10.2	11	18
Natural Gas	7.7	13	16
Coal	31.4	43.5	58
Total	49.2	67.5	92

*Low estimates represent current proven reserves plus likely reserves that can be recovered with existing technology at a reasonable price. These reserves have a high likelihood of recovery (greater than 90%)

** High estimates include the low estimates plus unconventional sources and additional undiscovered reserves that may exist but will require new technology and/or much higher market prices to make recovery economical. Achieving the high end of these estimates has a low likelihood of recovery (less than10%)

The geographic distribution of proven fossil fuel reserves is uneven. These are the reserves that are available to replace depleted fields in the next 30-50 years. As shown in Table 4.3, North America, Russia, and the Middle East are rich in proven oil, natural gas, and coal reserves, while most of the rest of the world has much less. North America has large reserves of unconventional oil and gas, as does Russia, South America, and the Middle East. China and Europe have intermediate reserves consisting mostly of sizable coal reserves. Africa and India have particularly limited reserves considering the population growth expected to occur by 2050.

Table 4.3. 2019 proven fossil fuel reserves at year-end

Region	Oil (Billion barrels)	Natural Gas (Trillion cubic feet)	Coal (Billion Tonnes)	Total ZJ*
North America	244.4	531.0	257.3	9.4
Russia/CIS	145.7	2266.8	190.7	8.9
Middle East	833.8	2670.0	1.6	8.0
China	26.2	296.6	141.6	4.6
Europe	14.4	118.7	135.1	4.0
Africa	125.7	527	13.2	1.7
South/Central America	324.1	282.1	13.7	2.7
India	4.7	46.9	105.9	3.2
Total World	1,733.9	7,019.0	1,069.6	49.1

* Assume 1 barrel of oil = 6 GJ; 1 mcf of natural gas = 1.1 GJ; 1 tonne of coal = 29 GJ

In 2019, the world consumed 98.3 million barrels of oil per day. Consumption grew by 1.3% per year from 2008 to 2018 and grew by 0.9% from 2018 to 2019. To meet expected demand, production in 20 years is going to have to increase to well over 120 million barrels per day. By 2035, today's existing fields will only be able to supply about 25-35 million barrels per day because these fields will have been depleted, especially with the use of modern recovery technology. Proven reserve oil fields that are currently undeveloped will provide substantial additional production, but unconventional oil and new undiscovered fields will be needed to meet demand. This will require substantial investment in oil exploration and production. Much of this additional capacity will be considerably more expensive to exploit. For example, profitably exploiting much of the existing oil sands and shale oil requires oil prices greater than $80 to $100 per barrel. Technological improvements are likely to drive this cost down, but the question is by how much? Many deep-water or arctic fields require even higher prices. Thus, over the long-term, oil is going to become more expensive and the risk to the environment to recover oil is going up.

Natural gas exploration and production has increased as a result of the development of modern recovery technology, including fracking. Current estimates of global total natural gas reserves (dry and wet) range from 13 to 16 ZJ of energy or 12 to 15 trillion mcf (mcf = thousand cubic feet). Total world proven reserves at the end of 2019 were 7 trillion mcf. The countries with the largest proven reserves were Russia (1,340 billion mcf), Iran (1,130 billion mcf), and Qatar (872 billion mcf). At the end of 2019, the US had 455 billion mcf of proven natural gas reserves. Africa, India, Europe, and China have much lower reserves. In 2019, world natural gas consumption was 138.8 billion mcf and production was 140.9 billion mcf. Natural gas currently accounts for about 24.2% of global energy production.

Natural gas has the lowest carbon footprint among fossil fuels. Burning coal to produce 1 million Btu's of energy produces about 206 to 229 pounds of carbon dioxide, depending on the type of coal burned. Burning fuel oil to produce the same amount of energy produces about 160 pounds of carbon dioxide, but burning natural gas produces only 117 pounds.[107] Thus, burning natural gas to produce electricity instead of coal reduces carbon dioxide emissions by almost 50%. In fact, much of the progress made by developed countries in meeting emission targets in the past 10 years has been achieved through replacing coal-fueled power plants with natural gas-fueled power plants.

Natural gas consumption is growing at about 2.0-2.5% per year in a global economy growing by 2.6% per year. If this growth rate were to continue through 2050, natural gas consumption would grow by 60-80% to about 250 billion mcf. At this rate of growth, total cumulative consumption from 2018 to 2050 would be more than 5 trillion mcf. If the growth rate in natural gas consumption should double to 3% per year as a result of accelerated replacement of coal with natural gas for power generation, total consumption through 2050 could reach 6.5 trillion mcf. If this later assumption were to be correct, by midcentury 40-50% of the world's total natural gas reserves and 80-85% of existing proven reserves could be depleted.

Natural gas power plants come in two varieties, simple cycle and combined cycle. They are also more efficient and lower cost to operate than coal or oil-fueled power plants. A combined cycle plant can be up to 60% efficient. A combined cycle plant costs about $500 to $550/kW (kilowatt) in the US. In 2019 world electricity production was about 27,000 terawatt hours (trillion watt hours) or 97 EJ. China and India accounted for 33% of global electricity production in 2019. If all this electricity were produced by natural gas, it would require burning about 0.2 to 0.25 trillion mcf per year. China and India do not have domestic natural gas proven reserves able to sustain this rate of consumption. China and India would need to import large amounts of natural gas from Russia, the Middle East, or the US. It would also require building about 14,000 to 15,000 250 MW

---

107 US Energy Information Administration, "Frequently Asked Questions: How Much Carbon Dioxide Is Produced When Different Fuels Are Burned?" (2020), https://www.eia.gov.

combined cycle power plants at a cost of over \$2 trillion. However, the conversion would reduce carbon dioxide emissions from electrical power generation in China by 30-40%.

Coal is the cheapest and most abundant fossil fuel, but it also produces the most carbon dioxide when burned to produce energy. In 2019 world coal production was 5.7 billion tonnes or 168 EJ of energy. Coal exploration and production has not grown recently because of the environmental impact of burning coal. If this trend continues, total consumption by 2050 will be about 4.5-5.5 ZJ or about 170 billion tonnes. There are, however, very large recoverable stores of coal on Earth. Estimates range from 1.5 to 2 trillion tonnes or about 43.5 to 58 ZJ of energy. Currently, proven reserves are about 1.07 trillion tonnes. Five countries have 75% of the world's proven coal reserves: US 23.7%, Russia 15.2%, Australia 14%, China 13.2%, and India 9.6%. The Middle East, Africa, and South America do not have large coal reserves.

The US Geological Survey (USGS) estimates undiscovered fossil fuel reserves based on the geology of a geographic region. Table 4.4 summarizes recent estimates.[108] The USGS uses a statistically based process to calculate the likely range of its estimates. The range of values extends from likely (greater than 95% likelihood of occurrence) to unlikely (less than 5% likelihood of occurrence). The estimate for US natural gas is a more recent estimate and is from the US EIA. Coal is not included in Table 4.4. Undiscovered coal reserves have been estimated to be about 800 billion tonnes and are likely located in the U.S., Russia, China, Europe, or Australia, which have over 80% of the Earth's known reserves.

Table 4.4: Undiscovered petroleum, natural gas, and natural gas liquids by geographic region reported by the US Geological Survey

Geographic Region	Petroleum (Billion barrels)	Natural Gas (Trillion mcf)
North America	25.5-208	2.0**
Middle East & North Africa	43-213	0.3-1.5
Former Soviet Union	16-166	0.3-3.3
Asia Pacific	21-88	0.3-1.1
Europe	4-19	0.05-0.25
Sub-Saharan Africa	41-232	0.3-1.2
South/Central America	45-262	0.1-0.8
South Asia	3-9	0.07-0.25

**Estimate from EIA in 2016

---

108 US Geological Survey, "World Oil and Gas Resource Assessments" (2020), https://www.usgs.gov.

If our needs through the end of the 21st century are ~75 ZJ and 85% of this energy, or 64 ZJ, is expected to come from fossil fuels, it can clearly be seen in Table 4.2 that even if oil and natural gas are much more abundant than currently estimated, oil and natural gas cannot provide all the energy humanity will need by 2100. Coal and alternative energy sources will have to make up the difference. If coal use is avoided because of its environmental impact, enormous amounts of alternative energy will need to be produced. Alternatively, technologies to capture and store or utilize carbon dioxide generated from burning coal will need to be deployed on a very large scale.

Alternative Energy Sources

There are several energy sources that are proven and reliable alternatives to burning fossil fuels. These alternative energy sources include solar, wind, nuclear, and geothermal energy. All of these sources have a much lower carbon footprint than fossil fuels. The efficiency and cost of producing energy from these sources has decreased substantially in the last several years due to engineering and other technical improvements. Around the world, new power plants using these technologies are being installed and brought online to help meet increasing demand for electricity and to reduce carbon emissions.

Not all of these sources are able to operate continuously throughout the year. The production capacity of each power plant is rated based on the amount of power produced at peak operation under ideal conditions. This rating is called the "nameplate" capacity. For example, a 200 MW plant is able to produce 200 MW/hour at peak operation and would produce 1,752 gigawatt-hours (GWh) of energy per year if the plant operated at peak production for every hour of every day of the year. Obviously, ideal conditions don't exist every hour of every day of every year for any of these technologies.

Measuring the actual power produced in a year and dividing by the nameplate capacity determines the "capacity factor" for a power plant. For example if a 200 MW plant actually produces 876 GWh in a year the capacity factor is 0.5 (i.e., 876 GWh produced/1752 GWh nameplate capacity). Solar and wind energy have capacity factors of 0.15 to 0.3 depending on local site-specific conditions, reflecting the fact that the sun shines less than 24 hours/day each day of the year and wind speeds vary from day to day. Geothermal power has a capacity factor of 0.75 to 0.85 and nuclear power plants have capacity factors of over 0.9. To put these capacities in perspective, the capacity factor for a US combined cycle natural gas power plant is 0.56.

Since power production from solar and wind energy is intermittent, energy may need to be stored in batteries to meet demand when energy isn't being produced. This requirement is especially important with small scale, local production when a solar installation or wind turbine is not connected to the power grid. When a solar or wind power plant is connected to a power grid that has other more persistent sources of energy also connected to it, fluctuations in power production can be easily managed and batteries are not required.

The cost of producing electrical power is an important consideration. The cost includes fixed costs such as the cost of construction, land, and financing the construction, and operating costs such as labor, raw materials, and maintenance. Some types of power plants have lower fixed costs and higher operating costs (e.g., coal) and some have higher fixed costs and lower operating costs (e.g., geothermal). Economists compare costs between different technologies using a measure called the "levelized cost of electricity (or energy)" or "LCOE." The LCOE is the net present value of all costs, fixed and operating, over the lifetime of a power plant, discounted for the cost of capital. LCOE can be weighted by capacity. Assumptions in these calculations include the expected life of the plant and the cost of capital, which can vary significantly between countries. The LCOE of a project does not capture all of the important cost considerations in comparing different technologies. Other cost considerations include the future costs that are avoided, which could include the cost of mitigating environmental degradation from fossil fuel plants.

Wind

Wind turbines have been developed that can operate with up to 40% efficiency in converting wind energy into electricity. Currently, most commercial wind turbines are 2 MW units that require about 0.75 acres per unit and cost $2.5-$3 million for each installed turbine. Turbines with a larger nameplate capacity have been developed, ranging up to 12 MW. The larger turbines are more cost effective. Installed costs range from $1,200/kW to $2,500/kW with a median of $1,600/kW. Annual operating costs are modest, and the turbines are expected to have a useful life of 15-20 years.

Offshore turbines are larger and more expensive to install per MWh. At the end of 2019, global installed nameplate capacity was about 650 GW and growing at about 60 GW per year. Wind turbines have the advantage of generating renewable energy that has a minimal carbon footprint and does not require freshwater. The disadvantages of wind turbine technology are the need to locate the turbines in geographic locations where environmental conditions generate substantial annual wind power, and the need to efficiently store, retrieve, and transmit the power when it is needed.

Most wind turbines start generating electricity at wind speeds of around 3-4 metres per second (8 miles per hour); generate maximum 'rated' power at around 15 metres per second (30mph); and shut down to prevent storm damage at 25 metres per second or above (50mph). The best places for wind farms are in coastal areas, at the tops of rounded hills, in open plains, or in gaps in mountains. Most regions of the world have wind power densities below 200 W/m². The ideal settings for wind farms have a wind power density of at least 800 W/m². Less than 10% of the area of the US has wind power density above 800 W/m², less than 6% of China, and less than 1% of India.

Wind turbines cannot be spaced closely because the rotors slow down the wind. The most efficient installations place the turbines between 7 and 15 rotor lengths apart and stagger their placement. The land requirement varies between 25 and 150 acres/MW, with a typical 200 MW installation requiring about 60 acres per megawatt nameplate capacity. The capacity factor for wind turbines is ~30%. A 200 MW wind farm would require about 40 5MW turbines costing about \$7.5-\$10 million each and 12,000 acres. The installation would generate about 500-600 GWh/year.

Solar

Solar energy can be generated by: 1) photovoltaic (PV) cells that directly convert solar energy into electrical energy; 2) solar-thermal energy in which solar energy is converted into heat that is used to produce steam that can drive turbines that generate electricity; and 3) solar fuels which are derived from plants or cyanobacteria (microalgae) that convert solar energy and carbon dioxide into biomass or hydrocarbons that can then be burned to produce energy.

The Earth receives about 7.7 ZJ of solar energy per day. Photovoltaic cells are currently about 20-25% efficient in capturing solar energy and converting it into electricity. Experimental PV cells can operate at 40-45% efficiency. Photosynthesis is about 8% efficient in converting solar energy into chemical energy. Thus, solar technology has enormous potential for delivering mankind's energy needs indefinitely. At the end of 2019 installed global capacity was about 630 GW and growing about 120 GW per year. Solar energy is reliable, low maintenance, and silent. The useful life of solar panels is about 30 years, after which performance drops below 80% of capacity.

The amount of solar radiation reaching the surface of a solar panel is a function of the latitude, time of the year, and the amount of solar radiation absorbed or reflected by the atmosphere and clouds. Lower latitudes and regions with fewer clouds and atmospheric particulates receive more solar radiation than higher latitudes or regions with extensive cloud cover. Most existing solar plants are located between 24-36°N latitude in arid regions with little cloud cover. In these regions the amount of solar radiation reaching the Earth's surface ranges from 2.5-7.0 kWh/m²/day, depending on the month, with an average of 5.75 kWh/m²/day. The land area required for a solar power plant depends on the technology used, the scale of production, and other site-specific factors. Currently installed 200 MW plants require about 3-5 km² (800-1200 acres). A 200 MW plant has about 700,000-750,000 solar panels with 1,140,000 m² of surface area. Panel efficiency varies from 15-20%. The productivity of the plant can be estimated with the following equation:

Incident solar radiation (kWh/m²/ day) x panel efficiency x surface area (m²) x 365 days/year = kWh/year.

For a 200 MW plant located in an arid region at about 35°N the equation is: 5.85 kWh/m²/ day x 0.2 x 1,140,000 m² x 365 days = 478.5 GWh/year. Nameplate capacity is 200MW x 8760 hours/year = 1,752 GWh/year, resulting in a capacity factor of 27.3%. This calculation does not account for lost capacity due to conversion from DC to AC power (which is ~5%). Capacity factors for solar power plants typically range from 15% to 30%.

To produce 0.1 ZJ/year or 2.778×10^{13} kWh/year it would require about 58,000 200 MW plants occupying 58 million acres. The cost to build a 200 MW plant varies by region and technology but a recent 200MW project due to be completed in the US in 2020 cost approximately $170 million. A 250 MW project in Chile in 2016 cost twice as much. To install 0.1 ZJ of energy production using $170 million 200 MW plants would cost ~$10 trillion.

Installed solar photovoltaic nameplate capacity at the end of 2018 was about 512 GW. At the end of 2019 nameplate capacity was 633.7 GW, which is equivalent to 5 EJ of delivered capacity at a capacity factor of 25%. World electricity production in 2019 was almost 100 EJ (27,000 TWh) so solar electricity production was about 5% of total global electricity production. China, Europe, and the US accounted for about 80% of wind production and about 55% of solar electricity production in 2019.

Nuclear

There are about 450 operating nuclear power plants in 30 countries and about 55 reactors under construction in 13 countries, most notably in China, India, UAE, and Russia. These power plants currently supply about 9.2 EJ (1.6%) of total world energy demand. Conventional nuclear power plants come in different sizes and designs and use uranium as a fuel. In these plants controlled nuclear fission produces heat that is used to make steam that drives turbines that make electricity. Uranium 235 (U_{235}) is used most commonly, although U_{238} and other actinium fuels can be used. The capacity factor for a nuclear power plant is approximately 0.9.

Uranium is plentiful on Earth. It is found in highly concentrated ores but can also be recovered in very small concentrations from seawater. The amount available is dependent on the cost of recovery and the extent that exploration has uncovered available supplies. As of 2019, there are about 6-8 million tonnes that can be recovered for less than $250,000/tonne. Annual consumption of uranium ore is about 86,000 tonnes per year. The current reserve to production ratio for uranium is about 90-100 years at uranium costs below $250,000 per tonne. The largest reserves are in Australia, Kazakhstan, Russia and Canada. Arctic and Antarctic reserves are unknown, but these reserves could be large and recoverable if ice caps, glaciers, and permafrost melt.

The outlook for the nuclear industry varies greatly depending on the expected life of existing reactors. Current installed capacity is 414 GW with new construction increasing that number to 440 GW in the next few years. However, there are about 25 GW of nameplate capacity scheduled for decommissioning by 2025 and another 55GW by 2030 based on a 60-year life expectancy. Most of these facilities are in Europe or Korea. Unless these facilities are re-licensed and reactors are able to operate safely for 80 years, about $30 billion/year will need to be invested just to replace aging facilities. The current rate of investment is about $10 billion per year.

There is a lot of complex and expensive science and engineering involved in the development and deployment of nuclear power. New designs and technology are continually being developed that could reduce the risks and costs of operation, but these advances are counterbalanced by low public acceptance of nuclear power and high levels of governmental regulation. Even though the safety record for the industry is good, high profile accidents and their serious and durable consequences have led some countries to significantly regulate or curtail the use of nuclear power. Other countries have continued to grow their capacity and deploy new technology, including China, Russia, and India. France derives in excess of 70% of its electrical power, about 380 TWh, from nuclear reactors while Germany has discontinued all nuclear power generation.

Nuclear power plants are expensive to build and require large amounts of natural resources to operate safely. The construction cost of a new nuclear reactor using U_{235} as fuel varies by country from about $2,000/kWh to about $7,000/kWh, with most countries paying overnight construction costs of $3,500 to $5,500/kWh. Thus a 1GW plant will cost $3.5 to $5.5 billion, plus other ancillary infrastructure costs such as the cost of land. Each plant requires about 8 square miles of land plus a variable exclusion zone for safety. Very large amounts of fresh water are needed for cooling. Closed-loop cooling systems use much less water than continuous flow cooling. A 1,000 MW plant with closed-loop cooling requires 70-80 million liters of fresh water per day, or about 7 billion gallons of fresh water per year.

Building a nuclear power plant takes about 6-12 years and a well-constructed modern plant can be expected to remain operational for 40-60 years. Obsolescence occurs because of the unavoidable impact of radiation on some essential components. Decommissioning a plant requires about 20 years, although nuclear waste from operations remains radioactive for thousands of years.

Operating costs are low and vary from country to country but are generally lower than alternative sources. The levelized cost of electricity (LCOE) takes the capital cost and operating cost into consideration. In 2025 the LCOE without tax adjustment for advanced nuclear power is expected to be $82 to $92 per MWh, compared to $38 to $45 per MWh for combined cycle natural gas, and $30 to $36 per MWh for solar photovoltaic panels.

Installation of approximately 0.1 ZJ of nuclear power capacity would require 3,500 1GW plants occupying about 17 million acres and costing about $13-$17 trillion. Of course, unless there are large undiscovered U_{235} reserves, there would not be adequate uranium fuel to operate even a fraction of this additional capacity at a reasonable cost. Without new technology to dramatically reduce the cost to recover U_{235} from unconventional sources, conventional nuclear power plants cannot increase the proportion of the world's energy demand they currently provide.

Nuclear power can also be generated by fast breeder reactors that use U_{238} as a fuel, which is about 100 times more abundant than U_{235}. These reactors produce large amounts of plutonium and can in some cases produce 30% more fuel than they consume. This plutonium can be reprocessed and used as a fissionable fuel itself. These reactors are more expensive to build and have technical challenges that make them much more risky and expensive to operate. They require use of liquid sodium, which is highly combustible, as a coolant. As of 2019, there are very few commercial breeder reactors in operation, and all are in Russia or China.

Nuclear fission offers mankind the opportunity to produce very large amounts of energy for a very long time without generating large amounts of greenhouse gases, but doing so will add very significant and durable risk and require enormous financial resources to fully exploit. As world energy demand will likely increase for the next 50-75 years, nuclear fission using U_{238}, reprocessed plutonium, or other isotopes could be an attractive alternative or may even become necessary. In the case of nuclear fission it's not about availability, it's about how much risk mankind is willing to take and how much of global GDP will be allocated to supply it.

Geothermal

Geothermal energy is generated using heat from the Earth. Energy from hot rock formations, hot springs, and magma can be used to produce steam that can then be used to drive turbines to produce electricity. Alternatively, the heat can be used directly. Geothermal resources can be found at or near the surface of the Earth (e.g., hot springs) or can be accessed by wells that are drilled into hot rock formations. Geothermal power plants have an average capacity factor of 0.75-0.85.

Direct use of thermal resources such as hot springs have been used for agriculture, cooking, and bathing for millennia. Geothermal power plants of various designs have been installed since about 1950. Installed nameplate capacity was 13.9 GW in 2019 and is growing at about 0.8 GW per year. North America and Europe account for about 55% of this installed capacity. The US expects to produce over 65TWh of electricity from geothermal power plants by 2050. Geothermal power plants do not require burning fossil fuels and can use briny or wastewater for steam production. Very little waste or emissions are produced. The largest geothermal installations are in the US (1.5 GW),

Mexico (0.72GW), and Italy (0.77 GW). Indonesia and the Philippines also have large installations. The typical cost of installation is \$2.5-\$5.0 million per MW, depending on site-specific expenses. In 2017 the weighted average installation cost was \$2959/kWh. The operating costs are very low, less than \$0.05 per kWh. The LCOE for geothermal energy is \$0.07/kWh (Range: \$0.03-\$0.12).

A 200 MW geothermal power plant will likely produce 1.314 GWh/year at a 0.75 capacity factor. It would take ~20,000 such plants to produce 0.1 ZJ of energy. A 200 MW geothermal plant costs about \$500-\$1,000 million to install but is very inexpensive to operate and only requires about 1-8 acres of land per MW.

Resource Requirements for Scale-up

Energy sources can be classified according to their carbon footprint. Burning fossil fuels or biomass has a high-carbon footprint whereas wind, solar, geothermal, nuclear and hydroelectric sources have a low-carbon footprint. In order for low-carbon footprint energy to supplant high-carbon footprint energy, low-carbon technology must be scaled up to produce zettajoules of energy over decades of service. In 2019 low-carbon energy sources supplied only 0.034-0.036 ZJ out of 0.584 ZJ or 6% of total world energy consumption. If we assume the nuclear power industry will only able to build new reactors fast enough to replace those that are becoming obsolete and hydroelectric production capacity will remain relatively constant, then new low-carbon power generation will have to come almost entirely from new solar, wind, and geothermal installations.

The resources required for production of 0.1 ZJ of energy from onshore wind turbines, solar photovoltaic panels, geothermal installations, and nuclear technology are shown in Tables 4.5 and 4.6.[109] Offshore wind turbines and solar thermal technology are more expensive and are not included in the table. Hydroelectric power, biomass, and hydrogen fuel cell technology are not scalable to a level of energy production equivalent to 0.1 ZJ and are not included. Nuclear and geothermal power plants have a much higher capacity factor than wind turbines or solar photovoltaic panels and so they are a more reliable source of energy. The cost of a commercial installation is about \$1,600-\$2,500/kW for wind, \$850-\$1,200/kW for solar, \$3,500/kW to \$5,500/kW for nuclear, and \$2,800/kW to \$3,000/kW for geothermal energy. Combining the expected capacity factor with the cost of construction, a solar "farm" has the lowest upfront capital cost per installed kW followed by a geothermal power plant.

Solar and wind power installations require much more land than nuclear and geothermal installations. Recently, interest has been growing in deploying both wind and solar offshore, even though capital costs are higher. Assuming there are no other limitations to

---

109 IRENA, "Renewable Power Generation Costs in 2019," International Renewable Energy Agency, Abu Dhabi (2020), https://www.irena.org/publications.

scale up, the cost of installing 0.1 ZJ of production is very large, at least $10 trillion for solar, and would require about 50 million acres (~200,000 km²), an area about half the size of California. Unfortunately, about 25-30% of this capacity would need to be replaced because of equipment obsolescence every 25-30 years. Fortunately, the levelized cost of electricity (LCOE) for wind ($0.06/kWh) and geothermal ($0.07/kWh) production is approaching the $0.046/kWh LCOE of energy produced by a modern commercial combined cycle natural gas power plants. The LCOE for solar power was about $0.07/kWh in 2019. However, the cost for solar PV technology has declined by more than 85% in the past 10-15 years and is expected to be less than $0.04/kWh in the near future.

According to the International Renewable Energy Agency, in 2019 the total nameplate deployment of solar, wind, and geothermal renewable energy sources was 1221.3 GW.[110] Adjusted for the respective capacity factors, this capacity would be expected to generate about 2,500-3,000 TWh of energy. If 2019 nuclear power generation of 2,796 TWh and hydroelectric power generation of 4,222 TWh are added; these low-carbon energy resources represent about 9,500-10,000 TWh of electricity generation, growing about 350-400 TWh per year. In 2019 these low-carbon sources represented about 36% of the 27,000 TWh of world electricity production but more than 85% of the growth in production capacity from 2018 to 2019.

In 2019 oil and natural gas exploration and new production investment was about $414.5 billion. Investment in utility scale renewable energy was $256.5 billion. This investment yielded about 110 GW of new solar nameplate capacity and about 50 GW of new wind nameplate capacity. In 2018 the total investment in wind was $127.5 billion for about 52 GW ($2,452/kW), of which $26.7 billion was for offshore wind. Investment was $130.8 billion for 109 GW ($1,200/kW) of solar utility-grade nameplate capacity. China invested $100 billion; Europe invested $75 billion; and the US $64 billion.

Installation of solar and wind electrical power generating capacity will continue to grow over the next 30 years as a result of their reduced installation costs and increased public demand. At the same time, some existing installations will reach the end of their useful life and new installations will just be replacing decommissioned installations. The rate of obsolescence is difficult to estimate, especially for new technology.

If we assume the net growth in low-carbon energy production capacity accelerates over the next 30 years to an average of 230 GW/year (70 GW wind and 160 GW solar), in 2050 low-carbon energy sources would be producing about 0.1 ZJ or about 28,000 TWh of energy. In 2050 world energy demand is expected to be 0.75-0.9 ZJ so low-carbon energy sources would represent about 11-13% of total energy demand. At the same time, world electricity demand is expected to be about 44,000-48,000 TWh so low carbon sources

---

110 https://www.irena.org.

would be supplying about 60% of world electricity demand. The cost of this deployment, assuming some further reduction in installation costs, would be about $140 billion/year for solar (in 2020 US$) and about $110 billion/year for wind production capacity for a total of $250 billion per year for 32 years or about $8.0 trillion through 2050.

Table 4.5: Comparison of alternative energy technology scalable to 0.1 ZJ/year of production.

Alternative Energy	Plant Nameplate Capacity	Capacity Factor	Number of Plants	Cost (Trillion $)	Land Requirement (Million Acres)
Onshore Wind	200 MW	0.3	53,000	16-20	600-700
Solar Photovoltaic	200 MW	0.2-0.25	58,000	8-10	50-75
Nuclear U$_{235}$	1GW Plant	>0.9	3,500	13-17	15-20
Geothermal	200 MW	>0.9	20,000	10-15	5-15

Table 4.6: Comparison of alternative energy technology scalable to 0.1 ZJ/year of production.

Alternative Energy	2019 US LCOE ($/kWh)	Unit Life (Years)	2018 Installed Nameplate Capacity (GW)	Annual Nameplate Capacity Growth (GW/year)
Onshore Wind	0.053	15-20	600	60
Solar Photovoltaic*	0.068	25-30	500	100
Nuclear U$_{235}$	0.075	40-60	363	12**
Geothermal	0.073	30-60	14.4	0.8

* Utility scale facility

** Nuclear installed capacity peaked in 2006 at 368 GW. At the end of 2018, 244 nuclear plants (60% of total) had operated for more than 30 years, including 77 that have operated for longer than 40 years. There are currently 50 reactors under construction of which 16 are in China. In the United States there are 35 reactors classified as uncompetitive and 6 slated for decommissioning.

Although growth in the capacity of the low-carbon energy sector (nuclear, hydroelectric, solar, wind, and geothermal) is increasing and over $330 billion was invested in 2018, the current trajectory is unlikely to result in low-carbon energy exceeding 15% of total energy demand in 2050. Furthermore, by 2050 obsolescence of existing installations will require a substantial annual investment in order to maintain production capacity.

Habitable Land

Humans need habitable land for agriculture; for villages, towns and cities; and for the infrastructure (e.g., lakes, reservoirs, etc.) necessary to support modern civilization. Agricultural land is used to grow crops or grow feed for livestock. Deserts, glaciers, mountains, forests, urban areas, and shrub or brush land are generally not suitable for growing crops or raising livestock to any significant degree. The total surface area of Earth is 196.9 million square miles (510.1 million km²). Oceans currently occupy 139.4 million square miles (361 million km²), and the total land surface area is 57.5 million square miles (149 million km² or 36.8 billion acres). Desert currently occupies about 33% of the total land surface area, mountains 20-24%, and forests 30-34%. Urbanization currently accounts for about 1.4 million square miles, but urban areas are expected to double in size in the next 30 years. Since some of these areas overlap (i.e., some mountains are covered by forest and some deserts have mountains), about 22-26% of the land surface area is available for agriculture or about 12.4±0.4 billion acres (48.6 to 51.8 million km²).

Agricultural land includes arable land that can be used to grow crops or used for meadows, and land that can be used as pastureland. Meadows are grassland used primarily for production of hay. The United Nations Food and Agricultural Organization (FAO) defines "Agricultural land" as the following:

1. Arable land, including cropland requiring annual planting and fallow land
2. Permanent cropland, including forested plantations (coffee, fruit, etc.)
3. Permanent pastures used for grazing livestock

Arable land can be cultivated, which means environmental temperatures, soil conditions, sunlight, nutrient availability, and water availability are suitable for growing crops. Land without adequate water for growing crops, with extreme temperatures, with soil that is rocky or land that is mountainous, or with soil that is polluted and/or nutrient poor is not arable land. Only about 10% of the Earth's total land surface area is arable, or about 3.5 billion acres (1.4 billion hectares).

The amount of arable land can be increased by deforestation, or irrigation and fertilization of non-arable land in regions that have suitable environmental and soil conditions, are not polluted, and are not rocky or mountainous. The amount of arable land can be decreased by urbanization, desertification, erosion of soil and nutrients, ocean level rise, and pollution. During the past 25 years large tracts of arable land have been lost to urbanization, erosion, and depletion of soil nutrients. This loss of acreage has been offset by deforestation and irrigation of non-arable land to use for growing crops. For example, in the last 20 years about

320 million acres of forest have been converted to agriculture, representing about 1% of the Earth's land surface area and about 3.2% of all forested areas. During this same time period, about 52 million acres of agricultural land was irrigated and fertilized to produce arable land. Worldwide fertilizer use has increased by about 1% each year for the last 20 years and is expected to continue to rise at this level for the foreseeable future. In contrast, about 30 million acres of arable land has been degraded each year for the past 20 years due to desertification, urbanization, pollution, or sea level rise. In summary, in the last 20 years about 370 million acres of arable land was added through deforestation and irrigation and about 360 million acres of arable land was lost due to desertification, urbanization, pollution or sea level rise. As a result, the total amount of arable land has not changed significantly in the last 20 years.

The amount of agricultural and arable land is not evenly distributed across the major geographic regions of the world. The United Nations Food and Agricultural Organization (FAO) maintains a high-resolution database (GLC-SHARE) that classifies land use into eleven different categories.[111] Data are gathered at the country level by the Land and Water Division of the FAO through local agencies or government agencies and integrated with satellite data. The database has a land cover spatial resolution of ~1 km² (247 acres). The map in Figure 4.3 uses the GLC-SHARE database to depict the type of land cover for each continent. The paucity of cropland (yellow) is evident as is the large amount of bare soil (white) in Africa, the Middle East, Asia, and Australia.

Figure 4.3. World land cover from the GLC-SHARE database

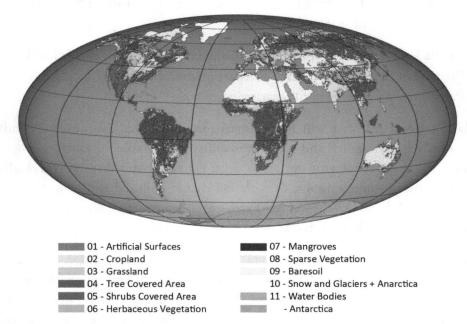

01 - Artificial Surfaces	07 - Mangroves
02 - Cropland	08 - Sparse Vegetation
03 - Grassland	09 - Baresoil
04 - Tree Covered Area	10 - Snow and Glaciers + Anarctica
05 - Shrubs Covered Area	11 - Water Bodies
06 - Herbaceous Vegetation	- Antarctica

111   John Latham, Renato Cumani, Ilaria Rosati, and Mario Bloise, "Global Land Cover (GLC-SHARE)," Beta-Release 1.0 Database, Land and Water Division (2014), http://www.fao.org/uploads/media/glc-share-doc.pdf.

Asia and Africa have the largest amount of agricultural land but have much less arable land per capita than Europe and North America. (Table 4.7) India has the largest amount of arable land with ~386.7 million acres (~156.5 million hectares). Russia and China each have ~296.5 million acres of arable land (~120 million hectares). By comparison the US has ~376.3 million acres (~152.3 million hectares) of arable land. Nigeria has 84 million acres (~34 million hectares) and is the country with the largest amount of arable land in Africa. Sudan (19.8 million hectares), Niger (16.8 million hectares), Ethiopia (15.1 million hectares), Tanzania (13.5 million hectares), and South Africa (12.5 million hectares) are the other African countries with the next largest amount of arable land.

Table 4.7. Amount of agricultural and arable land by geographic region

Geographic Region	Agricultural Land (Billion Acres)	Arable land (Billion Acres)	Arable land Per capita (Acres)	Total Land Area* (Billion Acres)
Asia	4.1	1.25	0.27	11.0
Africa	2.9	0.54	0.42	7.5
Latin America	1.8	0.37	0.56	4.5
Europe	1.2	0.69	0.93	2.5
North America	1.2	0.53	1.47	6.0
Oceania	1.1	0.11	2.75	2.1
Total*	12.3	3.49	0.45	33.6

*Does not include surface area of Antarctica = 3.2 billion acres.

The amount of agricultural or arable land needed in a country or region is determined by the size of the population and by their dietary practices. As per capita GDP and personal income increase, diets typically shift from those comprised of grains to diets that contain a greater proportion of meat, dairy, and eggs. As Africa and India grow in population and develop they are expected to experience an increase in per capita GDP. If this growth translates into increased personal income then the diet of these countries could change, especially their consumption of protein and meat. India may be the exception because of cultural and religious beliefs that promote a vegetarian diet.

The average human weighs about 137 pounds and needs to consume 1600 to 2300 calories per day to maintain their weight, depending on gender, age, and the amount of physical activity. The amount of dietary protein, carbohydrate, and fat varies greatly across the globe. In industrialized countries daily protein consumption is typically greater than 220 gms/day/person while in developing countries it is often less than 90 gms/day/person. The amount of land needed to generate sufficient calories varies greatly depending on the source of calories. Wheat, corn, and rice can generate 5 to 15 million calories per acre per year while most protein sources can only generate 1 million calories or less per acre per year. About 75% of all agricultural land (including crop and pastureland) is dedicated to animal protein production. The ratio of animal product calories to feed calories is, on average, only about 10% so using edible crops like corn to feed animals is an inefficient way to provide calories for humans.

The amount of land needed to meet dietary needs per capita per year varies depending on the amount of protein and fat consumed. In industrialized countries the amount of land required per person is about 0.8 to 1.5 acres (0.3 to 0.6 hectares) and in developing countries is about 0.4 to 0.8 acres (0.15 to 0.3 hectares). The amount of arable land needed varies from 0.25 to 0.75 acres per person per year (0.1 to 0.3 hectares), depending on agricultural productivity. Pastureland needs range from 0.25 acres to 1.5 acres (0.1 to 0.6 hectares), depending on the source and amount of dietary protein. The amount of land required assumes about 33% of the calories produced will be lost to waste during harvesting, storage, packing, transportation, sales, and cooking.

Improvements in agricultural productivity reduce the amount of arable land needed. There have been dramatic improvements in agricultural productivity over the last 50 years, driven by improved farming practices and technological advances in crop science and farm machinery. The world now uses about 70% less land to produce the same amount of food as it did in 1960.

Agricultural productivity is a function of latitude, water and fertilizer availability, agricultural practices, and technology used. For example, cereal yields in kilograms per hectare vary significantly between countries. In 2016 the US produced 8,143 kg/ha compared with 6,029 kg/ha in China, 4,181 kg/ha in Brazil, 2,650 kg/ha in Russia, 2,993 kg/ha in India, and 2,074 kg/ha in Australia. Not all agricultural products produced are used for food. About 55% of plant calories are used for human consumption. About 36% of plant calories are used as feed for livestock, which only returns about 4% as calories for human consumption. The remaining 9% of plant calories are used for industrial uses such as biofuels. In total about 41% of plant calories produced are lost for human consumption.

The amount of arable land needed per capita by geographic region is expected to decrease as a result of increases in agricultural productivity or changes in diet in most regions of the world. (Table 4.8) The amount of arable land needed per capita in Africa and Asia is expected to remain stable through 2050. Asia currently has sufficient arable land to be able to continue to support its population, especially in India. However, Asian countries will have to increase productivity or reduce waste if there is a change in dietary habits. There is not enough arable land in Asia and Africa to use arable land for industrial purposes. Africa does not have sufficient arable land to meet current dietary needs by 2050, even if the amount of arable land is increased by 50% due to deforestation or irrigation and fertilization of agricultural land.

Table 4.8: Estimated 2050 arable land needs

Region	2018 Arable Land (Billion Acres)	2018 Arable Land Per capita (Acres)	2050 Per Capita Arable Land Need (Acres)	2050 Estimated Population (Billion)	2050 Arable Land Needs (Billion Acres)
Asia	1.25	0.27	0.25	5.3	1.3
Africa	0.58	0.45	0.4	2.5	1.0
Europe	0.69	0.93	0.5	0.7	0.35
Latin America	0.37	0.56	0.5	0.8	0.4
North America	0.53	1.47	0.75	0.4	0.3
Oceania	0.11	2.75	0.5	0.06	0.03
World	3.49	0.45	0.38	9.76	3.38

The amount of agricultural land is sufficient to meet needs in 2050, although use of this land will need to be carefully managed to avoid ineffective or inefficient use. (Table 4.9)

By 2050, most of the arable and agricultural land currently available will be fully utilized. Some countries will depend on other countries or regions to meet their nutritional needs. Governments will need to ensure best agricultural practices are being used if they are to meet the dietary needs of their population, especially in Africa. Additional education, training, seed, and equipment is needed in order to succeed.

If world population reaches 11-13 billion in 2100 and the world has about 3.5 billion acres of arable land, only 0.286 acres of arable land and 0.735 acres of agricultural land are available per capita. Assuming that there will be significant improvements in agricultural productivity, changes in diet, new food production technology, and the current high level of ineffciency in world food production is improved through eliminating waste and creating an efficient system for moving food from productive to less productive regions of the world, the amount of agricultural land should be able to support the projected population. However, unless methods are developed to significantly expand the amount of arable or agricultural land by the second half of the 21st century, the Earth will most likely have reached its full capacity to support human habitation.

Table 4.9: Estimated 2050 pastureland land needs

Region	2018 Pastureland (Billion Acres)	2018 Pastureland Per capita (Acres)	2050 Pastureland Per Capita Need (Acres)	2050 Estimated Population (Billion)	2050 Pastureland Needs (Billion Acres)
Asia	2.85	0.63	0.4	5.3	2.1
Africa	2.36	1.8	0.6	2.5	1.5
Europe	0.51	0.7	0.7	0.7	0.5
Latin America	1.43	2.17	1.0	0.8	0.8
North America	0.67	1.86	1.0	0.4	0.4
Oceania	1.0	25	1.0	0.06	0.06
World	8.82	1.15	0.55	9.76	5.36

Fertile soils are limited and essentially non-renewable, at least on human time scales. It is crucial to protect available soil resources from degradation. The current amount of arable and agricultural land available for producing food should be sufficient to meet global demand through 2050, with the exception of Africa. Because of the expected increase in population in the next 30 years, the current amount of arable land in Africa is only about 50% of the arable land needed. To meet the shortfall African countries will experience a reduction in per capita calorie and protein consumption and will need to increase crop productivity, irrigate and fertilize pastureland to convert it into arable land, convert forested land to arable land, and import grain and cereals from regions of the world with excess production capacity. If climate change should reduce crop productivity, especially in Nigeria, Sudan, Ethiopia, and Tanzania, it is unlikely that sufficient arable land will be available to support the projected population growth in Africa.

Water

Modern civilization is dependent on water for commerce and activities of daily living. Most humans live on the shores of oceans and bays or the banks of rivers and major waterways because pumping water is energy intensive and expensive. Agriculture is dependent on the availability of freshwater and many industries, most notably electricity generation, require large amounts of water. Water is a finite natural resource. There are about 1.4 billion km³ of water on Earth. One cubic kilometer of water is equal to 264.2 billion gallons so there are about 370 million trillion gallons of water on Earth. Water is present as a liquid, a solid (ice and snow), or as a gas (water vapor). The natural reservoirs for water are oceans, seas, bays, ice caps, glaciers, snowfields, rivers, lakes, swamps, aquifers, groundwater, the biosphere, and the atmosphere. Oceans, seas, and bays contain the vast majority of the water on Earth, about 1.338 billion km³ of salt water. Saline groundwater adds another 12.9 million km³ and saline lakes 85,400 km³. The total amount of salt water on Earth is about 1.35 billion km³.

Ocean water has a salinity of about 3.5% while freshwater has a salinity of less than 0.05%. Saltwater is not suitable for agriculture or human consumption and is not used in most industrial settings because it is corrosive. Ice caps, glaciers and permanent snowfields contain about 29.2 million km³ of freshwater. About 10.5 million km³ of freshwater is ground water and the remainder is in lakes (91,000 km³), rivers (2,100 km³), and the atmosphere (12,900 km³). The freshwater total is about 39.8 million km³ or 10.5 million trillion gallons.

Water constantly moves between the natural reservoirs on Earth in a dynamic system called the water cycle.[112] (Figure 4.4) In the water cycle, water evaporates from oceans and from surface waters into the atmosphere. In addition, plants take up groundwater in their roots, and some of this water evaporates from leaves and flowers into the atmosphere in a process called transpiration. Some of the water vapor in the atmosphere condenses to form clouds. Clouds contain small liquid water droplets or particles of ice, which under the proper conditions produce precipitation (either rain, snow, or hail) that returns water to the surface of the Earth. Water on the surface can flow between reservoirs either as ice or liquid water, including rivers, glaciers, icebergs, and groundwater. Most groundwater is found within a half-mile of the Earth's surface. Water movement between land reservoirs can lead to runoff of surface or groundwater into the oceans. The water cycle provides about 45,000 km³ or 11,900 trillion gallons of renewable freshwater on land per year, net of evaporation from lakes and rivers and transpiration from plants.

Oceans cover 71% of the Earth's surface and play a pivotal role in the Earth's energy budget, and in its carbon and water cycles. Oceans are an important source of biodiversity, provide about 15-18% of the protein consumed by humans, and provide a means of transporting goods and raw materials over long distances at a reasonable price.

---

112 I. A. Shiklomanov, State Hydrological Institute (SHI, St. Petersburg) and United Nations Educational, Scientific and Cultural Organisation (UNESCO, Paris) 1999; Max Planck, Institute for Meterology, Hamburg. 1994; Freeze A., Cherry J. *Groundwater,* Prentice-Hall: 1979.

Humans are causing degradation of the world's oceans. Carbon dioxide emissions are causing acidification of seawater, global warming is causing ocean temperatures to rise, and glaciers and ice caps are melting, causing ocean salinity to decrease and sea levels to rise. Discharge of sewage, garbage, plastics, oil and petroleum products, and other forms of human waste is polluting the world's oceans at a high rate. The net effect of these physical effects is loss of coral reefs, loss of aquatic habitats, and reduction in aquatic biomass, including aquatic plant and animal life.

Figure 4.4: The Earth's water cycle. Numbers represent annual flux in km³.[113]

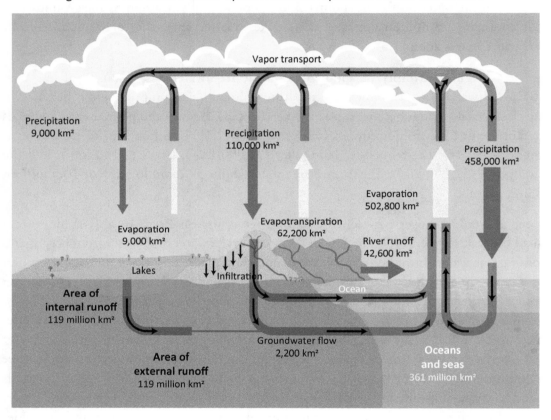

Freshwater resources on Earth total about 37-40 million km³ or about 2.85% of the total global water resources. Most of the freshwater is in glaciers and ice caps (currently about 28 million km³) or groundwater (10.5 million km³). The amount of groundwater varies by geographic region. Europe, Africa, and South America have the largest reserves of fresh groundwater.[114] (Figure 4.5)

113   https://serc.carleton.edu/eslabs/drought/1b.html

114   R. Taylor, B. Scanlon, P. Döll et al., "Ground Water and Climate Change," *Nature Clim Change* 3 (2013): 322–329, https://doi.org/10.1038/nclimate1744.

Figure 4.5. The world's major aquifers

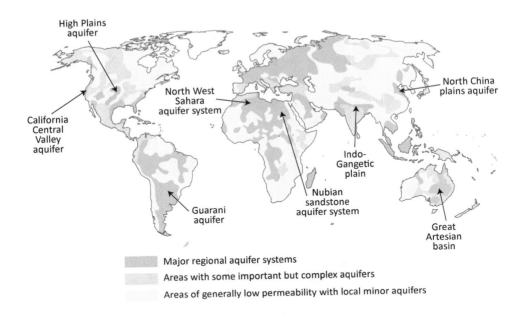

Major regional aquifer systems
Areas with some important but complex aquifers
Areas of generally low permeability with local minor aquifers

At any one time the atmosphere contains about 12,900 km³ of water, although annual global precipitation is about 510,000 km³ of which about 119,000 km³ falls over land. Ground ice and permafrost represents about 300,000 km³ while freshwater in lakes (91,000 km³), swamps (11,500 km3) and rivers (2,120 km³) contain the remainder of the fresh water on Earth. There are 263 major river basins on Earth that cover about 45% of the land surface area and contain 2,115 km³ of fresh water. Plant biomass contains a small amount of fresh water.

Freshwater in ice caps, glaciers, ground ice and permafrost, remote lakes and rivers, and swamps are inaccessible or do not provide ready access. Moving water any significant distance is energy intensive and expensive. Some aquifers lie deep in the Earth or are remote from civilization.

The FAO AQUASTAT database estimates the total amount of freshwater available for human consumption at about 43,000 km³ or only 0.1% of the freshwater on Earth. AQUASTAT is the FAO global information system on water resources and agricultural water management. It collects and analyzes data on freshwater production provided by country-specific governmental agencies. Humans withdraw about 4,500 km³ of freshwater per year from surface and groundwater for agriculture, industrial purposes, and domestic needs, including preparing food, personal hygiene, and drinking. Industrial uses are primarily related to electrical power generation.

Some of the water needed for agriculture comes from rainfall. Annual rainfall is distributed unevenly across the Earth and varies greatly by region and geography.[115] (Figure 4.6) Some areas are wet, and some are arid. About 4,760 km³ of annual rainfall (~4%) on land is used by agriculture. Annual rainfall on land is about 119,000 km³. About 62% of rainfall evaporates (~74,000 km³) and the remainder (~40,000 km³) finds its way into groundwater to replace the amount withdrawn or the amount that runs off into the oceans.

Figure 4.6: Average annual precipitation in cm/m²/year from 1960 to 1990. White is 0-1 cm/m²/yr; purple is 19-27 cm/m²/yr; dark purple is 99-247 cm/m²/yr.

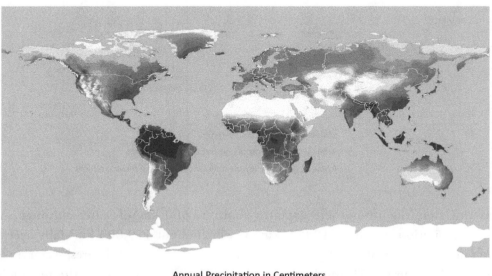

Annual Precipitation in Centimeters
0 4 7 10 14 19 27 38 99 247

Water for personal needs, industrial purposes, and for irrigation comes from freshwater withdrawal from lakes, rivers, and ground water. The amount of water available for withdrawal is important in those parts of the world where rainfall is not adequate for agriculture and domestic needs. Drinking and cooking needs are about 5-8 L (1.3-2.1 gallons) of potable water per day. In order to avoid dehydration humans need to drink 1-5 liters per day depending on body size, the ambient temperature and humidity, and the amount of physical activity. Domestic use of freshwater represents about 10% of total freshwater withdrawn worldwide. The amount of freshwater used for agriculture and by industry varies by country. North America and Europe use about 40-60% of freshwater for electrical power generation, 20-30% for irrigation, 5-10% for industrial purposes, and about 15% for domestic use. In Africa, 80-90% of freshwater withdrawn from ground or surface water is used for irrigation.

---

115   Center for Sustainability and the Global Environment, University of Wisconsin - Madison; Climate Research Unit, University of East Anglia. Retrieved from https://www.jpl.nasa.gov/edu/teach/activity/precipitation-towers-modeling-weather-data/

The amount of accessible, renewable freshwater available per capita can be used to determine the adequacy of freshwater resources. (Table 4.10) Water stress starts when the water available in a country or region drops below 1,700 m³/year per person. When the water available in a country or region drops below 1,000 m³/year per person, water scarcity is experienced. Absolute water scarcity is considered for countries with less 500 m³/year per person. Some 50 countries, with roughly a third of the world's population, experience water stress, and 17 of these extract more water annually than is recharged through their natural water cycles.

Table 4.10: Vulnerable regions are shown in yellow. Water stress is shown in orange and water scarcity in red.

	2014 Renewable Freshwater (Km3)	2017 Population (Million)	2017 Renewable Freshwater Per Capita (m3/person/year)
European Union	1,505	512.5	2,936.6
Russia	4,312	144.5	29,840.8
South America	13,868	644.1	21,530.8
North America	5,668	362.5	15,635.9
Sub-Saharan Africa	3,884	1,061.1	3,660.4
India	1,446	1,339.2	1079.7
Middle East & North Africa	231	444.3	519.9
East Asia	10,106	1,666	6,066
China	2,813	1386.4	2,029
South Asia	1,982	1,788.4	1,108.2
Australia	492	24.6	20,000
World	42,810	7,530.4	5,685

Groundwater provides about 90% of the freshwater withdrawals used each day in the world so the health of the groundwater resources is vital to meeting the needs of civilization. Groundwater resources in the southwest of the US, in the Middle East, and in central and south Asia are under stress.[116] (Figure 4.7)

---

116 T. Gleeson, Y. Wada, M. F. P. Bierkens, and L. P. H. van Beek, "Water Balance of Global Aquifers Revealed by Groundwater Footprint," *Nature* 448 (2012): 197-200.

Another approach to evaluating freshwater resources is based on water consumption rather than on water withdrawals from available resources. This method calculates the water footprint of a nation, business or industry. The water footprint is the total volume of freshwater used to produce the goods and services consumed by the nation, business, or industry.[117] The Water Footprint includes the amount of freshwater used in production and the amount of freshwater contaminated in the process. This approach differs from counting water withdrawals for domestic, industrial, or agricultural purposes.

The general approach is to determine how much water is withdrawn from surface or ground water for production (blue water), how much rainwater is consumed in production (green water), and how much freshwater is contaminated in the process (gray water). Data from the AQUASTAT database for the 10-year period from 1996 to 2005 has served as the basis for most of these analyses. There is uncertainty about the accuracy of some of the data and rainfall can vary greatly from year to year. It is assumed that using a 10-year time period will remove some of the variability related to the accuracy of reporting and variability in annual rainfall.

Figure 4.7. The world's ground water resources

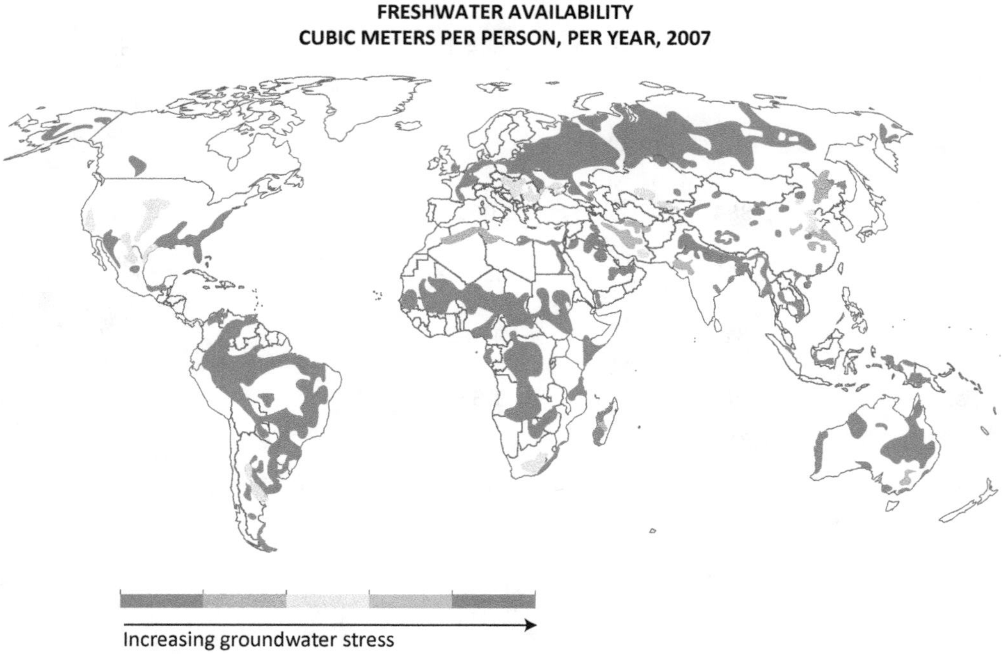

FRESHWATER AVAILABILITY
CUBIC METERS PER PERSON, PER YEAR, 2007

Increasing groundwater stress

From 1996 to 2005, total global water consumption was 9,087 km³/year of which 8,363 km³/year (92%) was used for agriculture. Interestingly, 1597 km³/year (19%) of the water used for agriculture was not used for domestic production; it was used to produce agricultural products

---

117   A. Y. Hoekstra and M. M. Mekonnen, "The Water Footprint of Humanity," *Proc. Natl. Acad. Sci.*, 109 (9) (2012): 3232-3237, https://doi.org/10.1073/pnas.1109936109.

for export. Green water (rainwater) accounted for 6,684 km³/year (80%) of the water used for agriculture. Blue water only accounted for 945 km³/year of the water used for agriculture. Industrial consumption was 400 km³/year (4.7%), and domestic consumption was 340 km³/year (3.8%). Blue water only accounted for 80 km³/year of the total 740 km³/year for industrial and domestic use. The remainder was gray water. The global per capita water footprint was 1,385 m³/year. Cereals consumption contributes the largest share to the global water footprint (27%), followed by meat (22%) and milk products (7%). Industrialized countries have per capita water footprints in the range of 1,250–2,850 m³/y. In developing countries the water footprint varies much more than for industrialized countries, ranging from 550–3,800 m³/y per capita. The difference between industrialized and developing countries is related to diet and meat consumption in particular. Gray water is largely fertilizer run off.

For governments in water-scarce countries such as in North Africa and the Middle East, it is crucial to recognize their dependency on external water resources and to develop foreign and trade policies such that they ensure a sustainable and secure import of water-intensive commodities that cannot be grown domestically. The water footprint of Chinese consumption is still relatively small and largely internal (90%), but given the country's rapid growth and the growing water stress (particularly in North China), the country is likely to increasingly rely on water resources outside its territory, evidenced by China's current policy to buy or lease lands in Africa to secure their food supply. A more detailed picture of current water scarcity is shown in Figure 4.8.

Figure 4.8. Fresh water availability per person per year

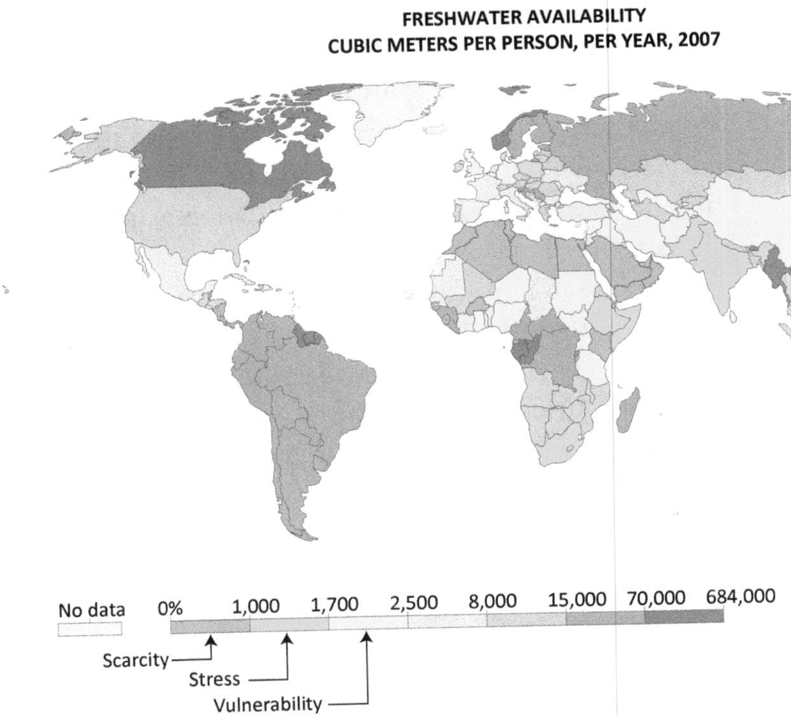

FRESHWATER AVAILABILITY
CUBIC METERS PER PERSON, PER YEAR, 2007

No data   0%   1,000   1,700   2,500   8,000   15,000   70,000   684,000

Scarcity
Stress
Vulnerability

Future Water Supply

Future supply of freshwater will be determined by the following factors:

1. Population growth
2. Global distribution of rainfall
3. Depletion of existing ground water resources
4. The regional or country-specific water footprint
5. The extent and pattern of international trade in products with a high water footprint

The effect of population growth is examined in Table 4.11. Based on population growth alone, India, the Middle East, and North Africa will experience water scarcity. Sub-Saharan Africa will experience water stress, which will be severe in some countries. China will have some regions experience water scarcity while other regions will experience water stress. The rest of the world would appear to have adequate water supplies.

The pattern of rainfall has changed in the last 10-15 years. Climatologists attribute this trend to changes in the Earth's climate due to greenhouse gas emissions. As predicted several years ago, clouds have moved to higher latitudes and become denser over lower latitudes. Regions in the mid-latitudes have experienced a reduction in precipitation and cloud cover.

Since 2002 NASA has measured the Earth's land-based water resources in its GRACE (Gravity Recovery and Climate Experiment) and Earth-observing satellite mission. The sum of groundwater, soil moisture, surface water, snow, and ice is measured as terrestrial water storage (TWS). Data gathered from 2002 until 2016 have been used to determine regional trends in TWS. TWS has decreased in Antarctica, Greenland, and the Gulf of Alaska due to rapid ice sheet and glacier melting due to climate change. TWS has increased in northern North America and Eurasia due to increased precipitation as predicted by several climate models. TWS is decreasing in many of the world's agricultural regions due to irrigation and ground water withdrawals, especially in Northern India and Northwest and North Central China. In general, the Middle East, South Asia, North Africa, and the southwest of North America are losing groundwater and surface water at an alarming rate.[118] These results suggest deserts in Africa, Asia, and North America are likely to get bigger and many regions of the world with marginal current supplies of freshwater are likely to experience scarcity by 2050. Should sub-Saharan Africa experience additional water scarcity; it is unlikely this region will be able to support population growth.

---

118   M. Rodell et al., "Emerging Trends in Global Freshwater Availability," *Nature* 557 (2018): 651-659.

Table 4.11: 2050 renewable freshwater per capita based on
2014 supply and projected 2050 population. .

	2014 Renewable Freshwater (km³)	2017 Population (Million)	2017 Renewable Freshwater Per Capita (m³/person/year)	2050 Renewable Freshwater Per Capita (m³/person/year)
European Union	1,505	512.5	2,936.6	3,344
Russia	4,312	144.5	29,840.8	33,426
South America	13,868	644.1	21,530.8	17,689
North America	5,668	362.5	15,635.9	13,090
Sub-Saharan Africa	3,884	1,061.1	3,660.4	1,618
India	1,446	1,339.2	1079.7	903
Middle East & North Africa	231	444.3	519.9	321
East Asia	10,106	1,666	6,066	6495
China	2,813	1386.4	2,029	2,000
South Asia	1,982	1,788.4	1,108.2	832
Australia	492	24.6	20,000	14,687
World	42,810	7,530.4	5,685	4,459

Summary

The natural world provides a variety of resources that are essential for mankind, including sunlight, fossil fuels, biomass, water, arable land, and minerals. Currently about 85% of the energy used by humans comes from burning fossil fuels such as oil, natural gas, and coal. Cheap, readily available energy makes modern agriculture, cities, and commerce possible.

World energy consumption is increasing due to population growth and continued economic development. Improvements in energy efficiency reduce world energy consumption.

North America, Russia, and the Middle East have large proven and unconventional oil, natural gas, and coal reserves, while most of the rest of the world, and especially Africa and India, has much less.

Since 1990 world energy consumption has grown on average about 1.6% per year and is currently about 0.59 ZJ/year. By 2050 world energy consumption is likely to be between 0.75 and 0.9 ZJ/year. From 2020 until 2050 the world will need to produce 20-22 ZJ of energy and by 2100 about 55-65 ZJ, assuming world energy efficiency improves by 35-40% over this time period. Total world recoverable reserves of oil are 11-18 ZJ and for natural gas are 13-16 ZJ. Total reserves of coal are 43.5 to 58 ZJ. Oil and natural gas cannot provide all the energy humans will need by 2100. Nuclear and hydroelectric energy provide 0.065 ZJ per year and are not expected to grow significantly. Renewables supply an additional 0.024 ZJ of energy.

To install approximately 0.1 ZJ/year of wind energy capacity requires about 53,000 200 MW onshore wind installations occupying about 600-700 million acres and costing $16-$20 trillion. Generating 0.1 ZJ/year of solar energy/year will require about 58,000 200 MW installations occupying 50-75 million acres of land and costing $8-$10 trillion. Installation of approximately 0.1 ZJ of nuclear power capacity would require 3,500 1GW plants occupying about 17 million acres and costing about $13-$17 trillion. Of course, unless there are large undiscovered U_{235} reserves, there would not be adequate uranium fuel to operate even a fraction of this additional capacity at a reasonable cost.

About 22-26% of the land surface area of Earth is available for agriculture or about 12.4 ± 0.4 billion acres. Arable land can be cultivated, which means environmental temperatures, soil conditions, sunlight, and nutrient and water availability are suitable for growing crops. Only about 6-10% of the Earth's total land surface area is arable, or about 3.5 billion acres. The total amount of arable land has not changed in the last 25 years. Asia and Africa have the largest amount of agricultural land while India has the largest amount of arable land, followed by the US and Russia. Asia and Africa have much less. The amount of agricultural or arable land needed in a country or region is determined by the size of the population and by the type and amount of food consumed. There have been dramatic improvements in agricultural productivity over the last 50 years, driven by improved farming practices and technological advances in crop science and farm machinery.

If world population reaches 11-13 billion in 2100 and the world has about 3.5 billion acres of arable land, only 0.286 acres of arable land and 0.735 acres of agricultural land will be available per capita. Assuming that there will be significant improvements in agricultural productivity, changes in diet, new food production technology, and reduction in the amount of food wasted, the current amount of arable land available should be sufficient to meet global demand through 2050, with the exception of Africa. If climate change should reduce crop productivity, especially in Nigeria, Sudan, Ethiopia, and Tanzania, sufficient arable land will not be available to support the projected population growth in Africa. If African arable land is used to produce food for export, food shortages in Africa in the next 30 years could reach catastrophic proportions. It is crucial to protect available soil resources from degradation, especially in Africa.

Water is a finite natural resource. There are about 1.4 billion km^3 of water on Earth. Water constantly moves between the natural reservoirs on Earth in a dynamic system called the water cycle. Oceans cover 71% of the Earth's surface, playing a pivotal role in the Earth's energy budget and its carbon and water cycles. Oceans, seas, and bays contain the vast majority of the water on Earth, about 1.338 billion km^3 of salt water. Humans are causing degradation of the world's oceans. The net effect of this pollution is loss of coral reefs, loss of aquatic habitats, and reduction in aquatic biomass, including aquatic plant and animal life.

The total amount of freshwater on Earth is about 39.8 million km^3. The global per capita water consumption was 1,385 m^3/year during the decade from 1996 to 2005, the last period for which these data are available. The amount of accessible, renewable freshwater available per capita can be used to determine the adequacy of freshwater resources. The total amount of renewable freshwater available for human consumption is only about 43,000 km^3 or about 11.4 million billion gallons.

Humans withdraw about 4,500 km^3 (1.2 million billion gallons) of freshwater per year from surface and groundwater for agriculture, industrial purposes, and domestic needs, including preparing food, personal hygiene, and drinking. About 4,760 km^3 (1.25 million billion gallons) of annual rainfall is used by agriculture. This amount represents about 4% of total annual rainfall. Groundwater provides about 90% of the freshwater withdrawals used each day in the world so the health of the groundwater resources is vital to meeting humanity's needs.

Groundwater resources in the southwest of the US, North Africa, the Middle East, and central and south Asia are under stress. Deserts in Africa, Asia, and North America are likely to get bigger and many regions of the world with marginal current supplies of freshwater are likely to experience scarcity by 2050. Based on population growth alone, India, the Middle East, and North Africa will continue to experience water scarcity. Sub-Saharan Africa will experience water stress, which will be severe in some countries. Should sub-Saharan Africa experience water scarcity this region will not be able to support projected population growth. Some regions in China will experience water scarcity while other regions will experience water stress. The rest of the world would appear to have adequate water supplies.

CHAPTER 5

Geography

"You can run, but you can't hide."
—Joe Louis

Chapter Guide

THIS CHAPTER DESCRIBES THE GEOGRAPHIC distribution of population growth, climate change, and natural resorce depletion. A risk register for each geograpghic region is presented along with a discussion of the data supporting the assessment. The stress tolerance of each region is evaluated using the World Bank's assessment of the rule of law, governance effectiveness, political stability, regulatory quality, and corruption control. Finally, the level of regional stress is related to the region's stress tolerance in order to evaluate the overall risk for adverse social, political, and economic consequences. A summary concludes the chapter.

Introduction

Population growth, climate change, and natural resource depletion are likely to produce significant social, political, and economic consequences that will differ by region or country. Countries vary in their demographic, cultural, and sociopolitical profile; the resources they have to deal with challenges; and in the effectiveness of government. The degree to which people will be adversely affected by the consequences of population growth, climate change, and natural resource depletion is a function of the environmental stress produced and the tolerance a country or region has for dealing with stress. Those countries or regions with high stress tolerance or those that are less affected will fare better than those that have low stress tolerance or who are likely to be severely affected. In order to evaluate the regional impact of population growth, climate change, and natural resource depletion a list of outcomes was developed to gauge risk. The time horizon was 2100 unless specifically noted otherwise. The outcomes evaluated and the impact assessments are shown in Table 5.1.

Table 5.1: Impact of population growth, climate change, and natural resource depletion

Impact	Very High	High	Moderate	Low
Population Growth*				
Population Growth	>50%	>20%	≥10%	<10%
Population Decline	>50%	>20%	≥10%	<10%
Population >65 years old	>35%	>25%	>15%	≤15%
Immigration per year	>5 MM or 5%	>1 MM or 1%	>0.5 MM or 0.5%	≤0.5 MM or <0.5%
Emigration per year	>5 MM	>1 MM	>0.5 MM	<0.5 MM
Per Capita GDP growth	<0%	<1.5%	<2.5%	≥2.5%
Cities ≥30 million in 2050	>5	>2	>1	None
Climate Change				
Sea Level Rise >1 meter	Subjective classification based on multiple factors			
Desertification				
Violent Storms				
Natural Resource Depletion				
Freshwater per capita	<500 m^3	<1000 m^3	<1700 m^3	≥1700 m^3
2018 Arable land/ 2050 Arable land need	<0.5	<1.0	<1.25	≥1.25
2050 Energy Reserve/ Production Ratio	<10	<20	<30	>30

*MM means million and % refers to the percent of the total population or % change in per capita GDP.

The impact of rising sea level, desertification, and violent storms was subjective based on assessment of the outcomes that would likely be experienced in each region. The likelihood of occurrence was also evaluated. Low likelihood was <10% chance of occurrence; moderate was 10% to 20%; high was 21% to 50%; and very high was >50%. The regions evaluated are shown in Table 5.2.

Table 5.2: Regions evaluated.

Region	Main countries included	Comment
South Asia	India, Pakistan, Bangladesh	Includes Afghanistan
East Asia	China, Japan	Includes Korea and Taiwan
Southeast Asia	Indonesia, Thailand, Philippines, Vietnam	Includes Malaysia
North America	US, Canada	Excludes Mexico
Latin America	Brazil, Mexico, Argentina	Includes Central America
Russia		
Middle East	Saudi Arabia, Egypt, Turkey, Iran	Excludes Armenia, Azerbaijan, Georgia
Africa	Nigeria, Congo, Ethiopia	Excludes Egypt and Madagascar
Australia		
Europe	Germany, UK, France	Includes Switzerland and Norway

South Asia

Population Stress

- India is the largest country in the region with 77% of the population. South Asia is expected to grow from a population of 1.87 billion in 2017 to a population of ~2.5 billion by 2100. India's population should stabilize by mid-century while the population in rest of the region will continue to increase.
- Like the rest of the world, the region's population will age but less than 15% of the population is expected to be over age 65 by 2050.
- The region has been a source of human capital and intellectual talent for other regions of the world and will likely continue to do so. However, those leaving the country permanently represent a small proportion (~0.1%) of the population. Remittances from the Indian diaspora are 2.5-3.0% of GDP.
- Hindus are the largest religious group in India. Muslims are ~15% of the population in India, ~96% in Pakistan, and 90% in Bangladesh. The total fertility rate (TFR) among Muslims in the region is about twice the rate for Hindus and population momentum is greater for Muslims. In India the TFR for Muslims was 2.6 in 2016 and for Hindus was 2.1 live births per woman of childbearing potential. Both

rates have decreased by about 20% in the last decade. Despite decreasing TFR for both populations, population momentum will result in a Muslim population in India in 2050 of 350-375 million, up from 200 million in 2016.

- Economic growth is projected to be high (3-4%), although underemployment and income disparity will continue or increase.
- There are four cities projected to have more than 30 million inhabitants in 2050: Mumbai, Delhi, Kolkata, and Karachi. These same cities are projected to have more than 50 million inhabitants by 2100.

Climate Stress

- The Ganges-Brahmaputra Delta is Asia's largest. Sea-level rise is projected to displace millions of people in this delta and increase the salinity of groundwater, endangering agricultural productivity.
- The Thar Desert is in northwest India and extends into Pakistan. This region would be at risk of expanding if the climate warms and precipitation shifts to higher latitudes.
- India has an active monsoon season and is vulnerable to severe thunderstorms and dust storms.
- Temperatures in South Asia are likely to exceed the habitable range in the late 21st century.[119]

Natural Resource Stress

- The region currently experiences freshwater stress and is projected to experience significant freshwater scarcity as the population grows. The scarcity could become absolute and a major problem if the Ganges River and aquifer system become further depleted because of loss of Himalayan glaciers. South Asia gets >70% of its rainfall from the monsoons. Any sustained reduction in the amount of rainfall from the monsoons would have a devastating effect on freshwater resources in South Asia.
- India has the largest amount of arable land in the world but only about 0.28 acres/capita. The amount of arable land could be significantly reduced if the Ganges-Brahmaputra Delta is inundated. By 2050, the amount of arable land per capita is expected to drop below the amount needed to fully support the population, especially if groundwater is in limited supply.
- India has large reserves of coal and is heavily dependent on coal for its future energy supply. Oil and natural gas reserves cannot support future demand. In 2017 India's oil reserve to production ratio was only 14.4 years. Nuclear and

---

119 E.-S. Im, J. S. Pal, and E. A. B. Eltahir, "Deadly Heat Waves Projected in the Densely Populated Agricultural Regions of South Asia," *Science Advances* 2 (2017): e1603322, http://advances.sciencemag.org.

renewables provide <5% of energy consumption. India will have to import energy and/or massively invest in nuclear or renewable power plants to meet demand.

The risk register for South Asia is shown in Table 5.3.

Table 5.3: South Asia Risk Register

Risk	Impact	Likelihood	Risk
Population Growth			
Population Growth			High
Population Age >65			Low
Loss of human capital			Low
Per Capita GDP growth			Low
Cities > 30 MM			High
Climate Change			
Sea Level Rise >1 meter			High
Desertification			Moderate
Violent Storms			High
Natural Resource Depletion			
Freshwater Per Capita			High
Arable Land			High
Energy			High

East Asia

Population Stress

- China is the largest country in the region with 85% of the population. The population of East Asia is expected to be about 1.2 billion by 2100, which is a reduction of about 400 million people from its peak of about 1.6 billion in about 2030. Like the rest of the world, the region will age.
- In 2050 more than 35% of Japanese and more than 25% of Chinese will be over the age of 65. Corresponding percentages for age 80 are 15% and 7.5%. This segment of the population will place a significant burden on these economies.
- The region has been a major source of human capital for other regions of the world and will continue to do so, but emigration and significant loss of human capital is not a major risk given the size of the population.
- Tokyo and Shanghai are projected to have more than 30 million inhabitants in 2050. Continued growth beyond this level is unlikely.
- Economic growth is projected to remain high. Income disparity will continue to grow.

Climate Stress

- East Asia would be severely affected by a sea level rise of 1 meter or more. More than 32,000 square kilometers of China's coastal area and more than 23 million people are at risk. Japan is an island country, and Korea is a peninsula. Inundation from cyclonic storm surges will intensify.
- The Gobi Desert occupies 1.3 million km² of northern and northwestern China and parts of Mongolia and is at risk of expanding if precipitation shifts to higher latitudes. Korea and Japan do not have significant deserts.
- East Asia is particularly vulnerable to severe storms. Cyclones strike Japan, Korea, and the eastern coast of China and are capable of affecting millions of people living within a few kilometers of the coast.

Natural Resource Stress

- The region is currently vulnerable to freshwater stress and could experience additional stress if the climate warms.
- China currently has 0.23 acres of arable land per capita; the Koreas have 0.12 acres per capita, and Japan has only 0.08 acres per capita. These levels are already well below levels needed to support the population. The projected decrease in population by 2100 should reduce stress related to the amount of arable land per capita. The amount or quality of arable land could be significantly reduced if coastal regions are inundated due to sea level rise or if ground water reserves decrease or there is saltwater incursion into aquifers.
- China has large reserves of coal and is heavily dependent on coal for its future energy supply. Oil and natural gas reserves cannot support future demand. Total consumption for the region was 0.16 ZJ in 2017, representing about 27.6% of total world consumption. Less than 6% of production was from nuclear or renewable energy. Only China has meaningful fossil fuel reserves, which total about 5.5 ZJ. Although large these reserves could provide energy for more than 30 years at projected future consumption levels. Reserve to production ratios are currently 18.3 years for oil, 36.7 years for natural gas, and 39 years for coal. Nuclear and renewables provide only about 6% of energy consumption.

The risk register for East Asia is shown in Table 5.4.

Table 5.4: East Asia risk register

Risk	Impact	Likelihood	Risk
Population Growth			
Population Decline			High
Population Age >65			High
Loss of human capital			Low
Per Capital GDP growth			Low
Cities > 30 MM			Moderate
Climate Change			
Sea Level Rise >1 meter			High
Desertification			Moderate
Violent Storms			High
Natural Resource Depletion			
Freshwater Per Capita			Moderate
Arable Land			Very High
Energy			High

Southeast Asia

Population Stress

- Southeast Asia includes Indonesia, Thailand, the Philippines, Vietnam, Malaysia, Myanmar, Cambodia, Laos, Brunei, Singapore, and East Timor. The 2017 population was about 650 million. Indonesia is the largest country in the region with 41% of the population. The Philippines is second with 16%. The population of Southeast Asia is expected to increase by about 19% to 800 million by 2100. The region is not expected to have any significant population growth after about 2050.
- In 2050 only 13% of Indonesians will be over the age of 65.
- The region, and especially the Philippines, has been a source of human capital for other regions of the world and will likely continue to be so.
- Economic growth is projected to be high (2.5-4.2%) in the region, led by Thailand (>3%) and Vietnam (>4%).
- Urbanization has progressed in the region but there are no cities projected to

have more than 30 million inhabitants by 2050 or even 2100. By 2030 ~60% of the population will be living in cities; up from 45% in 2008. Almost 30% of the urban population lives in slums.

Climate Stress

- A sea level rise of 1 meter or more would be catastrophic for much of Southeast Asia. Indonesia is an island country and has 54,700 kilometers of coastline. The Mekong delta encompasses 40,500 km^2 in Southwest Vietnam. One third of Thailand's rivers flow into the Mekong River. The Mekong delta is a fertile, flat flood plain. Much of the agricultural production in the region occurs in heavily populated low-lying regions near the coasts. Coastal erosion and seawater intrusion could reduce rice yields by 50% or more.
- Southeast Asia does not have any deserts and desertification is unlikely.
- Southeast Asia is highly vulnerable to severe storms. More than 20 typhoons strike the region each year. These storms are increasing in intensity, cause significant economic damage, and affect the lives of millions of people living within a few kilometers of the coast.

Natural Resource Stress

- Most countries in the region currently have more than 3,000 m^2 freshwater per capita and receive large amounts of annual rainfall. Indonesia has more than 7,500 m^2 of freshwater per capita. Surface water availability in Vietnam and Thailand has decreased because of extensive construction of upstream dams on the Mekong River in China, reducing downstream flow. Groundwater resources are decreasing in Thailand and Cambodia.
- Southeast Asia currently has 0.34 acres of arable land per capita. Indonesia has 0.34 acres per capita, the Philippines have 0.25 acres per capita, and Thailand has 0.64 acres per capita. The projected stabilization of the population by 2050 should prevent stress related to the amount of arable land per capita from developing based on population growth. The amount and/or the quality of arable land could be significantly reduced if coastal regions are inundated due to sea level rise or if ground water reserves continue to decrease.
- Southeast Asia does not have large reserves of fossil fuels and produces less than 1% of its energy from renewable sources. Regional consumption is about 0.02 ZJ per year. Current proven reserves cannot support future demand. In 2017 Reserve to production ratios for oil and coal are less than 20. The region does not produce nuclear energy.

The Southeast Asia risk register is shown in Table 5.5.

Table 5.5: Southeast Asia risk register

Risk	Impact	Likelihood	Risk
Population Growth			
Population Growth			Moderate
Population Age >65			Low
Loss of human capital			Moderate
Per Capital GDP growth			Low
Cities > 30 MM			Low
Climate Change			
Sea Level Rise >1 meter			High
Desertification			Low
Violent Storms			High
Natural Resource Depletion			
Freshwater Per Capita			Low
Arable Land			Moderate
Energy			High

North America

Population Stress

- The current population of North America is 0.36 billion. The population of North America is expected to be 0.4 billion by 2100, an increase of 10%. Population growth will be driven by immigration since there is no significant population momentum in North America and the current regional total fertility rate is 1.8 births per woman of childbearing potential, which is below replacement levels. If immigration into the region is greater than 1 million per year, then the population could reach 0.5 billion or more by 2100, a 39% increase.

- Like the rest of the world, the region will age. In 2050 more than 20% of Americans and more than 25% of Canadians will be over the age of 65. Corresponding percentages for age 80 are 8% and 11%. This segment of the population will be an economic burden unless people over 65 years of age remain in the workforce and/or are able maintain their health later in life.

- Immigration has been and will continue to be a major source of population growth in North America. In 2017 there were about 44.5 million immigrants residing in the US, representing about 12.5% of the US population and growing at a rate of 3% per year. In 2017 about 44% of immigrants were from Latin America, and 27% were from Asia. There are more than 8 million immigrants in Canada representing about 22% of the population.

- Per capita GDP growth in North America is expected to be 1-1.5% through 2050. National debt obligations and trade deficits will reduce GDP growth in the US.
- Urbanization will continue in North America, but no cities are projected to have more than 30 million inhabitants by 2100.

Climate Stress

- North America will be adversely affected if sea levels rise one meter or more. The US has more than 95,000 miles of shoreline. Florida and tidal wetlands are at highest risk. Coastal cities will need to build barriers to prevent water incursion.
- North America has three large deserts in the southwest, extending into Mexico. Collectively these deserts cover more than 700,000 km². Groundwater in these desert regions is being depleted. Expansion of these deserts is likely to occur if precipitation patterns shift to higher latitudes. Expansion of these deserts would reduce the amount of arable land in southern California and in central Texas.
- North America is vulnerable to severe storms. Hurricanes and tornadoes routinely strike central and southern North America and are capable of affecting millions of people. Intensification of these storms could result in widespread loss of property and life.

Natural Resource Stress

- North America has ample freshwater reserves, especially in Canada. However, freshwater resources are not evenly distributed. The Southwest of the US will experience water scarcity, including southern California. Groundwater reserves are decreasing in the Northwest coastal regions of Canada and Alaska. Rainfall and groundwater resources should increase in northern US states and in Central and Eastern Canada.
- North America currently has 1.2 billion acres of agricultural land and 0.53 billion acres of arable land (1.47 acres per capita). These levels are above levels needed to support the population. By 2100, the amount of arable land will exceed the amount needed based on current population projections. The amount or quality of arable land could increase in northern North America if global temperatures rise, precipitation patterns move north, and northern latitudes have prolonged growing seasons or become arable.
- North America has large proven reserves of fossil fuels, on the order of about 11 ZJ. Consumption in 2017 was 0.108 ZJ. Primary energy consumption has not increased significantly in the last 20 years and varies by less than 0.5% per year. The reserve to production ratio for proven reserves is >35 years for oil, >11 years for natural gas, and >335 years for coal. These ratios do not include additional unconventional and undiscovered reserves. Nuclear and renewable energy provide about 12% of consumption. Fracking and enhanced oil recovery (EOR) have greatly increased production capacity in North America.

The North American risk register is shown in Table 5.6.

Table 5.6: North America risk register

Risk	Impact	Likelihood	Risk
Population Growth			
Population Growth			Moderate
Population Age >65			Moderate
Immigration			Moderate
Cities > 30 MM			Low
Per Capital GDP			High
Climate Change			
Sea Level Rise >1 meter			High
Desertification			Moderate
Violent Storms			High
Natural Resource Depletion			
Freshwater Per Capita			Low
Arable Land			Low
Energy			Low

Latin America

Population Stress

- Latin America includes Mexico and the countries of Central and South America. Brazil is the largest country with 33% of the population followed by Mexico with 20%. The current population of Latin America is approximately 650 million. The population of Latin America is expected to peak at about 800 million in mid-century and stabilize at 700-750 million by 2100. The fertility rate for the region was 2.06 in 2017 (Range: 1.7-2.4) with positive population momentum.
- By 2050 more than 20% of the population will be over the age of 65. This segment of the population will be an economic burden unless people over 65 years of age remain in the workforce and/or are able maintain their health later in life.
- The emigration rate from Latin America has decreased in the last decade, especially from Mexico. However, from 2010-2017 the emigrant population grew by about 2.5 million people, mostly from Central America, Venezuela, and the Dominican Republic, and now represents about 10% of the total Latin American population. Economic instability and violence drive continued emigration. The United States

is the destination for about 65% of Latin American emigrants. The loss of skilled labor through emigration is offset by remittances so the regional impact is neutral.
- Per capita GDP growth in Latin America is expected to be 2.6-3.4% through 2050 led by Brazil, Mexico, Argentina, and Columbia.
- Urbanization will continue in Latin America. Mexico City will have more than 30 million inhabitants by 2100. More than 20% of the total population of Mexico lives in Mexico City.

Climate Stress

- Latin America will be adversely affected if sea levels rise by a meter or more. The Yucatan Peninsula, Central America, the Caribbean, and coastal cities are at greatest risk. Some Caribbean islands would be inundated by a sea level rise of this magnitude, displacing millions of people.
- Latin America has multiple large deserts. The Sonora and Chihuahua deserts cover about 600,000 km² of Mexico's landmass. The Atacama, Patagonian, and Sechara deserts cover more than 900,000 km² of South America. Groundwater in these desert regions is being depleted, especially in Mexico and Patagonia. Expansion of these deserts is likely to occur if precipitation patterns shift to higher latitudes.
- Mexico, Central America, and the Caribbean are vulnerable to severe storms. Hurricanes routinely strike coastal areas and Caribbean islands and are capable of affecting millions of people. Intensification of these storms would result in widespread loss of property and life.

Natural Resource Stress

- South America has ample freshwater reserves, especially in Brazil. However, freshwater resources are not evenly distributed across Latin America. Mexico and parts of Central America currently experience freshwater stress and are vulnerable to scarcity. Freshwater scarcity is likely to worsen in Mexico and could become absolute if precipitation patterns change and rainfall moves to northern latitudes.
- Latin America currently has 1.2 billion acres of agricultural land and 0.37 billion acres of arable land (0.56 acres per capita). These levels should be adequate to support the population through 2100.
- Latin America has large proven reserves of oil and natural gas. Total reserves are about 2.8 ZJ. Consumption in 2017 was 0.03 ZJ. Primary energy consumption has grown by about 2.1% per year for the last 10 years. The reserve to production ratio for proven reserves is 126 years for oil, 46 years for natural gas, and 141 years for coal. These ratios do not include additional unconventional and undiscovered reserves. The distribution of oil, natural gas, and coal in Latin America is very uneven; 90% of oil reserves and 80% of natural gas reserves are in Venezuela. Brazil has 50% of the region's coal. Nuclear and renewable energy provide only about 5% of consumption.

The risk register for Latin America is shown in Table 5.7.

Table 5.7: Latin America risk register

Risk	Impact	Likelihood	Risk
Population Growth			
Population Growth			High
Population Age >65			Moderate
Loss of human capital			Moderate
Cities > 30 MM			Low
Per Capital GDP			Low
Climate Change			
Sea Level Rise >1 meter			High
Desertification			Moderate
Violent Storms			High
Natural Resource Depletion			
Freshwater Per Capita			Low
Arable Land			Low
Energy			Low

Europe

Population Stress

- Europe includes the countries of the European Union (EU) plus Albania, Belarus, Bosnia, Kosovo, Macedonia, Moldova, Norway, Switzerland, and the Ukraine. It does not include Russia, which is treated separately. Germany is the largest country but has only about 13.5% of the population. The current population of Europe is about 605 million, not counting Russia. The population of Europe is expected to decrease by about 10% by the end of the century. The total fertility rate was 1.59 in 2017 (Range: 1.3-1.9) with negative population momentum.
- The region's population is aging. By 2050 more than 25% of the population will be over the age of 65 (Range: 24-36%). This segment of the population will be a significant economic burden unless people over 65 years of age remain in the workforce and/or are able maintain their health later in life.
- Immigration has increased significantly in the past 10 years, especially Moslems from Africa and the Middle East. In 2017 2.4 million immigrants from non-EU countries entered the EU bringing the number of non-EU citizens up to 22.3 million. The EU naturalized 825,000 citizens in 2017. By 2050 Moslems could represent 15% or more of the EU population.

- Per capita GDP growth in Europe is expected to be 1.5-2.0% through 2050.
- Over 70% of Europeans currently live in urban areas. Further urbanization is unlikely.

Climate Stress

- Europe will be adversely affected if sea levels rise by a meter or more. Coastal cities and agricultural areas are at greatest risk from saltwater incursion or inundation.
- Desertification and severe storms do not appear to be major risks for Europe.

Natural Resource Stress

- Europe has ample freshwater reserves, especially given the decline in population expected through 2100. Surface and groundwater resources could increase if precipitation patterns shift to more northern latitudes.
- Europe currently has 1.2 billion acres of agricultural land and 0.69 billion acres of arable land (0.93 acres per capita). These levels are adequate to support the population through 2100.
- Europe does not have large proven reserves of oil and natural gas. Total proven reserves, including coal, are about 3.9ZJ. Consumption in 2017 was 0.08 ZJ. Primary energy consumption has decreased slightly by about 0.6% per year for the last 10 years presumably reflecting improved energy efficiency. The reserve to production ratio for proven reserves is 10.4 years for oil, 12.2 years for natural gas, and 159 years for coal. Europe represents less than 10% of the world's annual coal consumption, which partially explains the high reserve to production ratio. Nuclear, hydroelectric, and renewable energy provided about 0.02 ZJ (25%) of consumption in 2017 and is increasing about 5% (0.001 ZJ) per year. Europe will be increasingly dependent on foreign sources of energy unless nuclear and/or renewable energy grows at a much greater rate or coal reserves are used to a greater degree.

The risk register for Europe is shown in Table 5.8.

Russia

Population Stress

- The current population of Russia is about 143.4 million. The population of Russia is expected to decrease by about 15-20% by the end of the century. The current total fertility rate is 1.8 and declining with negative population momentum.
- Russia's population is aging. By 2050 more than 20% of the population will be over the age of 65. This segment of the population will be an economic burden unless people over 65 years of age remain in the workforce and/or are able maintain their health later in life.
- Reliable data on migration into and out of Russia is unavailable. There are likely to be a large number of illegal immigrants living temporarily in the country

(estimates range from 4-9 million), many of whom are from former Soviet states. These transient immigrants are employed, move into and out of the country periodically, and will likely continue to do so in the future.

- Per capita GDP growth in Russia is expected to be about 1.5-2.0% through 2050.
- Internal migration from rural areas to cities will continue but there are no cities projected to have greater than 30 million inhabitants this century.

Climate Stress

- It is unlikely a sea level rise will have a major effect on Russia. Most major urban centers are located inland. Thawing of the arctic and tundra will likely increase Russia's energy and agricultural resources.
- Desertification is not a major risk for Russia.
- Severe storms causing widespread personal and property damage are uncommon in Russia.

Table 5.8: European risk register

Risk	Impact	Likelihood	Risk
Population Growth			
Population Decline			Moderate
Population Age >65			High
Immigration			High
Cities > 30 MM			Low
Per Capital GDP			High
Climate Change			
Sea Level Rise >1 meter			Moderate
Desertification			Low
Violent Storms			Moderate
Natural Resource Depletion			
Freshwater Per Capita			Low
Arable Land			Low
Energy			High

Natural Resource Stress

- Russia has ample freshwater reserves, especially given the projected decline in population. Surface and groundwater resources could increase if precipitation patterns shift to more northern latitudes.

- Russia currently has 0.55 billion acres of agricultural land and 0.3 billion acres of arable land (2.1 acres per capita). Climate change could significantly increase these resources.

- Russia has very large proven reserves of oil and natural gas. Total proven reserves, including coal, are about 7.8 ZJ. Consumption in 2017 was 0.03 ZJ. Primary energy consumption has grown slightly by about 0.3% per year for the last 10 years. The reserve to production ratio for proven reserves is 25.8 years for oil, 55 years for natural gas, and 391 years for coal. Coal represents only about 10% of Russia's energy consumption, which partially explains the high reserve to production ratio. Nuclear and hydroelectric energy provided less than 10% of consumption in 2017. There is very little renewable energy capacity in Russia. Russia exports the majority of its energy production and can be expected to continue to do so through the end of the century. Oil and natural gas revenues accounted for 36% of Russia's federal budget revenues in 2016.

The risk register for Russia is shown in Table 5.9.

Table 5.9: Russia risk register

Risk	Impact	Likelihood	Risk
Population Growth			
Population Decline			High
Population Age >65			High
Immigration			Moderate
Cities > 30 MM			Low
Per Capital GDP			Moderate
Climate Change			
Sea Level Rise >1 meter			Low
Desertification			Low
Violent Storms			Low
Natural Resource Depletion			
Freshwater Per Capita			Low
Arable Land			Low
Energy			Low

Middle East

Population Stress

- The Middle East includes Saudi Arabia, Egypt, Yemen, Oman, Iran, Iraq, Kuwait,

Israel, Qatar, The United Arab Emirates, Syria, Lebanon, Turkey, Palestine, Jordan, Bahrain, and Cyprus. In 2017 the population was about 400 million. Egypt, Iran and Turkey are the largest countries in the region with about 60% of the population. The population of the Middle East is expected to increase by about 75% to 700 million by 2100. Four countries are expected to provide more than 85% of the population growth for the region through 2100, Egypt, Yemen, Iraq, and Syria.

- In 2050 less than 15% of the population will be over the age of 65.
- The region has been a source of human capital for other regions of the world and will likely continue to do so. In 2017 more than 18 million migrants from the region worked in Europe or within other countries in the region. Syria, Palestine, Iraq, and Turkey were the largest source of migrants.
- Growth in per capita GDP through 2050 is expected to be 1.5-2.5% in the region, depending on cash flow from oil and natural gas markets, the status of economic sanctions, and the existence of military conflict. Qatar, Kuwait, UAE, and Saudi Arabia have some of the highest per capita GDP's in the world, driven by oil and natural gas wealth and a low population. Syria, Yemen, and Iraq are among the poorest countries in the world because of military conflicts. Iran's economy has been affected by periodic US economic sanctions.
- Urbanization has progressed in the region but there are no cities projected to have more than 30 million inhabitants by 2100.

Climate Stress

- A sea level rise of 1 meter or more would seriously threaten coastal cities in the Middle East. Coastal cities such as Doha, Dubai, and Abu Dhabi are located in low-lying coastal zones or on islands. As many as 24 port cities are at risk for inundation from sea level rise.
- Deserts dominate the landscape in the Middle East. Desertification of marginal land will reduce agricultural productivity and GDP growth, impoverish rural inhabitants, and lead to further urbanization. Fertile regions in Turkey, Iran, and Syria are at risk from global warming, especially if groundwater resources are depleted.

Natural Resource Stress

- The Middle East has less than 300 m³ of renewable freshwater per capita. Turkey and Iran have the largest freshwater reserves and are not vulnerable to water scarcity. Most countries in the region have much less than 500 m³ per capita of renewable freshwater reserves and receive less than 10 inches of annual rainfall. Desalination is a major source of fresh water but requires a large initial capital investment and large amounts of energy to maintain operations.
- The Middle East currently has ~0.2 acres of arable land per capita. Iran has 0.45

acres per capita, Iraq has 0.32 acres per capita, and Saudi Arabia has 0.26 acres per capita. Most of the arable land is near the coast or major rivers. Over 80% of freshwater withdrawals are for agriculture. The amount and/or the quality of arable land could be significantly reduced if coastal regions are inundated due to sea level rise or if ground water reserves continue to decrease.

- The Middle East has very large reserves of fossil fuels. Oil and natural gas reserves amount to nearly 8 ZJ. Regional consumption is about 0.037 ZJ per year. Nuclear energy provides about 1% of the regions energy. Nuclear energy provides ~20% of Israel's energy consumption and represents about 60% of the regions installed capacity. Renewable energy provides less than 0.2% of energy consumption in the Middle East, although renewable energy is expected to grow in importance over the next decade.

The risk register for the Middle East is shown in Table 5.10.

Table 5.10: Middle East risk register

Risk	Impact	Likelihood	Risk
Population Growth			
Population Growth			Very High
Population Age >65			Low
Loss of human capital			Moderate
Per Capital GDP growth			Moderate
Cities > 30 MM			Low
Climate Change			
Sea Level Rise >1 meter			High
Desertification			High
Violent Storms			Low
Natural Resource Depletion			
Freshwater Per Capita			Very High
Arable Land			High
Energy			Low

Africa

Population Stress

- The continent of Africa has 54 countries and is often divided geographically into North Africa and Sub-Saharan Africa. North Africa consists of Morocco, Libya, Tunisia, Egypt, Algeria, and Sudan. Nigeria, Ethiopia, and Egypt are the most populated countries in Africa. The current population of Africa is about 1.3 billion. The population of Africa is expected to increase to 4.4 billion by the end of the century. The total fertility rate in North Africa was 2.7 in 2017 (Range: 2.2-3.2) and in Sub-Saharan Africa was 4.85 live births per woman of childbearing potential combined with positive population momentum.

- Africa's population is young; less than 3.5% of the population is over age 65. Over 40% are under the age of 15. By 2100 about 40% of the working age population of the world will be from Africa.
- The majority of migration in Africa is within the region to neighboring countries. Emigration from Africa and especially from Sub-Saharan Africa has grown in the last decade. About 17-20 million Africans live outside of Africa. Most African migrants come from North Africa. The Middle East, Europe and the US are the main destinations. Emigration is caused by unemployment, low wages, conflict, and political instability.
- Per capita GDP growth in Africa is expected to be 1.5-2.5% through 2050. Most African countries have emerging economies. Performance can vary widely based on multiple social, political, and economic factors. Climate change, the ability to shift from agrarian to industrial economies, the availability of capital, and the level of education are just a few of the factors that will influence future growth.
- Urbanization and desertification are major risks for Africa. Migration from rural to urban areas will create 5 of 10 largest cities in the world by 2050. By 2100 these cities will likely have more than 50 million inhabitants. The rate and extent of growth will likely be greater than the ability of these cities to build adequate housing and infrastructure. Many people will live in slums and will not have access to the essentials of life. Unemployment or underemployment rates will be high.

Climate Stress

- Africa will be adversely affected if sea levels rise by a meter or more. Coastal cities and agricultural areas are at greatest risk from saltwater incursion or inundation.
- Africa has three major deserts, the Kalahari, the Namib, and the Sahara. The total area of these deserts is 2.48 billion acres of which the Sahara is 2.25 billion acres. The Sahara covers North Africa and the other two are located in West Africa. Expansion of these deserts would seriously threaten cropland and grassland in Nigeria, and West Africa.
- Hurricanes and severe storms do not result in large annual losses of life or property in Africa.

Natural Resource Stress

- North Africa has absolute water scarcity with most countries having less than 400 m^3 per capita of renewable fresh water. Sub-Saharan Africa is vulnerable to water stress, depending on the country. Population and economic growth will place greater stress on freshwater resources in Sub-Saharan Africa and could lead to persistent stress or scarcity.

- Africa currently has over 2.5 billion acres of agricultural land and 0.58 billion acres of arable land (0.45 acres per capita). These levels should be adequate to support the current population. If population growth meets expectations, the efficiency of agricultural practices will need to dramatically increase to meet demand.
- Africa does not have large proven reserves of oil and natural gas. Total proven reserves, including coal, are about 3.9ZJ. Consumption in 2017 was 0.08 ZJ. Primary energy consumption has increased by about 3% per year for the last 10 years. The reserve to production ratio for proven reserves is 42.9 years for oil, 61.4 years for natural gas, and 53 years for coal. Nuclear, hydroelectric, and renewable energy provided less than 1% of consumption in 2017. Hydroelectric provides about 0.6%. Only South Africa has a nuclear power capacity. Population and GDP growth will place an increasing burden on Africa's energy resources. These resources will not last until the end of the century.

The risk register for Africa is shown in Table 5.11.

Table 5.11: African risk register

Risk	Impact	Likelihood	Risk
Population Growth			
Population Growth			Very High
Population Age >65			Low
Loss of human capital			Moderate
Per Capital GDP growth			Moderate
Cities > 30 MM			Very High
Climate Change			
Sea Level Rise >1 meter			Low
Desertification			Very High
Violent Storms			Low
Natural Resource Depletion			
Freshwater Per Capita			Very High
Arable Land			High
Energy			Low

Australia

Population Stress

- The current population of Australia is about 24.6 million. The Australian population is expected to grow by over 70% by the end of the century to more than 40 million. The total fertility rate was 1.8 in 2017 with neutral population momentum. The population growth rate has been stable at 1.5-1.7% per year for the past few years. About 55%-65% of this growth is due to migration.
- The region's population is aging. By 2050 more than 20% of the population will be over the age of 65 and more than 5% will be over age 80.
- Over 25% of the population was foreign born in 2016. In the last 10 years about 200,000 to 250,000 immigrants entered Australia per year, net of migrants leaving the country. Only about 15-20% of migrants have permanent Visas. After the United Kingdom and New Zealand, the largest immigrant populations are from China, India, the Philippines, and Vietnam.
- Per capita GDP growth in Australia is expected to be 1.3%-1.5% through 2050.
- Urbanization does not appear to be major risk for Australia through 2100.

Climate Stress

- Australia will be adversely affected if sea levels rise by a meter or more. Coastal wetlands in the Northern Territory are at greatest risk from saltwater incursion or inundation.
- At least 20% of the land area of Australia is desert. The Great Victoria Desert and the Great Sandy Desert are the largest. However, as much as 35% of the land area is arid. Expansion of Australian deserts due to global warming could put arable land at risk, particularly in the South.
- More than 85% of Australians live within 50 km of the coast. The El Niño Southern Oscillation heavily influences Australian weather. An increase in severe storms, combined with sea level rise, would put coastal regions at high risk of salt-water incursion, erosion, and flooding. Typhoons strike the northern coast and especially the northwestern coast of Australia.

Natural Resource Stress

- Australia has ample freshwater reserves that should be adequate through 2100 even with the projected growth in population. Rainfall patterns in Australia are variable putting some marginal arid regions at risk. Groundwater resources are increasing, especially in the Northeast. Australia is exporting large amounts of fresh water as agricultural products. High water footprint products such as beef, cotton, and dairy products represent more than 30% of Australia's agricultural exports.

- Australia currently has 0.97 billion acres of agricultural land and 0.11 billion acres of arable land (4.6 acres per capita). These levels will support the population through 2100. Agriculture represents about 3 percent of the Australian economy. More than 75% of agricultural production is exported.
- Australia has large proven reserves of coal but very little oil and natural gas. Total proven reserves, including coal, are ~5.5 ZJ. Consumption in 2017 was 0.006 ZJ. Primary energy consumption has increased by about 1.1% per year for the last 10 years. Over 70% of Australia's energy is produced from coal and about 15% from natural gas. The reserve to production ratio for proven reserves is 31.6 years for oil, 32 years for natural gas, and 301 years for coal. Hydroelectric and renewable energy provided only about 6% of energy consumption in 2017. Australia has never had a nuclear power plant although Australia has 33% of the world's uranium deposits.

The risk register for Australia is shown in Table 5.12.

Table 5.12: Australian risk register

Risk	Impact	Likelihood	Risk
Population Growth			
Population Growth			Very High
Population Age >65			Moderate
Immigration			Moderate
Cities > 30 MM			Low
Per Capital GDP			High
Climate Change			
Sea Level Rise >1 meter			High
Desertification			High
Violent Storms			Moderate
Natural Resource Depletion			
Freshwater Per Capita			Low
Arable Land			Low
Energy			Low

Population Risk

The size, structure, and distribution of human populations will change dramatically in the 21st century. The regional stress associated with population growth is shown in Table 5.13.

Table 5.13: Regional population stress. (See Table 5.1 for color code)

Region	Population Stress					
	Population Growth/ Decline	Immigration/ Emigration	Urbanization	Aging Population	Slow Economic Growth	Overall Population Stress
South Asia	Growth					
East Asia	Decline					
Southeast Asia	Growth	Emigration				
North America	Growth	Immigration				
Latin America	Growth	Emigration				
Russia	Decline	Immigration				
Middle East	Growth	Emigration				
Africa	Growth	Emigration				
Australia	Growth	Immigration				
Europe	Decline	Immigration				

World population will increase from 6.1 billion in 2000 to 11.2 billion (Range: 10-13 billion) in 2100. Population growth will be uneven; Sub-Saharan Africa, the Middle East, and South Asia will account for over 90% of this growth. Africa will account for over 75% of population growth. There are just 10 countries that will generate more than 50% of this population growth (Table 14). The population of East Asia, Russia, and Europe will decline, resulting in a reduction in the number of people in the workforce, aging of the population, and reduced economic growth.

Table 5.14: The ten countries with the largest increase in population from 2000 to 2100

Country	Population (Million)		
	2000	2100 Projection	Increase
Nigeria	122.9	752.2	629.3
India	1,053.5	1659.8	606.3
Democratic Republic of the Congo	48.0	388.7	340.7
Pakistan	138.3	364.3	226.0
Uganda	23.8	202.9	179.1
Ethiopia	66.4	242.6	176.2
Iraq	23.6	163.9	140.3
Egypt	68.3	200.8	132.5
Kenya	31.1	156.9	125.8
Sudan	28.1	127.3	99.2
Total	1,604.0	4,259.4	2,655.4

Population growth will be accompanied by intra-regional migration from rural to urban centers as people seek economic opportunity and improved living conditions. By 2100 there will likely be 20 cities with more than 50 million inhabitants; all of these cities will be in Africa and South Asia. Dense population will likely place a heavy burden on infrastructure and local government.

If the need for housing, water, and public health is not met, large areas of substandard housing without essential utilities such as electricity, sewage and water management could become even more common. Many countries in the developing world will likely have more than 30% of their population living in slums. Access to basic healthcare, including reproductive healthcare, will continue to be very limited. Unemployment or underemployment could exceed 50% unless means to provide economic opportunity are developed.

These cities could become breeding grounds for disease, crime, and opportunists, and enclaves in some of the largest cities could be ungovernable. Lawful economic opportunity for many will likely be unavailable or nonexistent. The value of human capital will decrease. Desperation could force many to seek a better life through any available means. Alternatively, national and local governments may provide necessary resources and services, and mechanisms will be developed to provide economic opportunity and basic healthcare, including birth control. If governments take these actions, then the adverse consequences of urbanization can be mitigated.

The number of the world's transnational migrants will increase to more than 600 million by 2100. Migration will increase due to economic and environmental deprivation, political repression, and regional conflict or violence. Africa, the Middle East, and South Asia will be the source of these migrants who will continue to seek residence in North America, Europe, and Australia. Southeast Asia will continue to be a source of migrants. Russia will likely experience increased immigration. East Asia appears to be at low risk for a substantial increase in immigration. In each of these regions, immigration will provide a source of young labor to increase GDP growth, but the expected number of migrants is very large. In those regions that will experience a large influx of migrants, social and cultural conflicts will emerge that could become serious if immigrant assimilation is not managed effectively.

Between 2010 and 2030 the world's population will likely grow from 6.9 billion to 8.3-8.5 billion. Currently the population growth rate for Muslims is twice the growth rate for non-Muslims. This difference is due to a higher total fertility rate and positive population momentum. As a result, by 2030 between 600-650 million of the additional 1.4-1.6 billion people on Earth since 2010 will be Muslim. India will have the largest Muslim population in the world and over 15% of Europeans will be Muslim by 2050. All of the 20 countries with the largest population growth rate are predominantly Muslim, including countries outside of Africa.

These data show that world population growth can be largely explained by an African effect and a Muslim effect. It will not be possible to reduce world population growth without understanding the cause of these disparate population growth rates and learning how to control them. If the current demographic trends persist, by 2100 over 40% of all people on Earth of working age between 15 and 65 years will be African Muslims.

Climate Risk

By 2100 global warming due to anthropogenic carbon emissions will lead to profound climate changes that will not be evenly distributed by region (Table 5.15). Most regions of the world will experience high levels of stress due to environmental degradation. This stress will cause major social, economic, and political consequences. Southeast Asia and Oceania will experience the most stress from sea level rise while desertification will cause the most stress in the Middle East and Africa. A shift of agricultural zones to higher latitudes could reduce global agricultural productivity and lead to famine in Africa, South Asia, Central America, and the Middle East. Most regions are at high risk from an increase in violent storms such as typhoons, hurricanes, and tornadoes. Russia will experience a low overall level of climate stress because of its northern latitude and the relative paucity of coastal cities or arable land at risk.

Table 5.15: Regional climate stress. (See Table 5.1 for color code)

Region	Climate Stress			
	Sea Level Rise >1 meter	Desertification	Violent Storms	Climate Stress
South Asia				
East Asia				
Southeast Asia				
North America				
Latin America				
Russia				
Middle East				
Africa				
Australia				
Europe				

Natural Resource Risk

The world's natural resources are not evenly distributed. (Table 5.16) The adequacy of existing resources is a function of the current reserves and the rate at which the reserves will be used through the end of the century. The stress level from depletion of freshwater resources is especially high in South Asia, Africa, and the Middle East. These same regions are at risk of not having adequate arable land to support the expected population growth, especially if climate change should shift precipitation to higher latitudes and cause desertification.

Asia and Europe do not have adequate energy reserves. China and India are especially dependent on coal for electrical power. While these countries can be expected to pursue alternative energy sources, the cost and time needed for these efforts will lead to continued high levels of carbon emissions for at least the next few decades. North America, Latin America, Russia, and Australia are endowed with large natural resource reserves. These regions will have to supply Asia, Europe, the Middle East, and Africa, directly or indirectly, in order to avoid natural resource scarcity contributing to transnational migration or promoting international conflict.

Table 5.16: Regional natural resource stress. (See Table 5.1 for color code)

Region	Freshwater per Capita	Arable Land	Energy	Natural Resource Stress
South Asia				
East Asia				
Southeast Asia				
North America				
Latin America				
Russia				
Middle East				
Africa				
Australia				
Europe				

Political Systems

Forming governments and establishing laws, rules, and regulations are mechanisms through which most societies attempt to ensure some level of safety and security, provide basic services, adjudicate disputes, and promote economic activity. Effective governance is essential to managing risk and promoting regional cooperation. The World Bank has established a method for measuring governance based on five characteristics:

1. Rule of Law
2. Governance effectiveness
3. Political stability
4. Regulatory Quality
5. Control of corruption

The World Bank uses multiple data sources, undertakes a thorough assessment of their validity and accuracy, and then collates the information into a percentile ranking from 0 (lacking) to 100 (ideal) for each of these elements. The current percentile ranking for selected countries is shown in Table 5.17.

Table 5.17: Governance effectiveness and stress tolerance

Region or Country	Rule of Law	Governance Effectiveness	Political Stability	Regulatory Quality	Corruption Control	Stress Tolerance
Top Five Countries by 2050 Population						
China	45	68	37	49	47	Low
India	53	57	17	42	42	Low
US	92	93	59	93	89	High
Indonesia	41	55	29	52	48	Low
Brazil	44	42	31	51	36	Low
Second Five Countries by 2050 Population						
Pakistan	24	31	2	29	23	Very Low
Nigeria	19	16	5	17	13	Very Low
Bangladesh	28	22	10	11	19	Very Low
Russia	22	50	21	33	17	Very Low
Mexico	32	52	23	10	16	Very Low
Selected Other Populous Countries						
Germany	91	94	67	95	92	High
Japan	90	93	89	90	90	High
Kenya	38	41	13	44	15	Low
Ethiopia	34	24	8	14	33	Very Low
Australia	93	92	78	91	92	High
Egypt	33	29	9	17	34	Low
Iran	26	45	16	10	20	Very Low
Saudi Arabia	57	63	24	55	66	Moderate
UK	93	91	57	94	95	High
South Africa	52	65	36	63	57	Moderate
Thailand	55	67	19	60	43	Moderate
Vietnam	56	53	60	37	32	Moderate

These rankings have been used to assess the ability of countries to tolerate stress due to environmental degradation, population growth and migration, and natural resource depletion. Stress tolerance is defined as "very low" if three or more of the scores are less than 25; stress tolerance is "low" if three or more of the scores are less than 50; stress tolerance is "moderate" if three or more of the scores are less than 75; and stress tolerance is "high" if three or more of the scores are ≥75. These data highlight the variability of stress tolerance between countries and demonstrate that some of the regions that are expected to encounter the greatest environmental, population, or natural resource stress, such as Africa, the Middle East, and Pakistan, are likely to have the least stress tolerance.

Global Risk

The level of stress tolerance has been compared with the risk profile for each region in order to assess overall risk for adverse social, political, and economic consequences from population growth, climate change, and natural resource depletion. (Table 5.18) All regions of the world face risks; they just differ from region to region. Unfortunately, the regions with the highest risk have the lowest stress tolerance. The regions with the highest stress tolerance face some of the lowest risks, which are largely related to high immigration pressure and risks related to climate change.

Table 5.18: Global total risk

Region	Total Risk				
	Population	Climate	Resources	Stress Tolerance	Total Risk
South Asia	Increase				
East Asia	Decline				
Southeast Asia					
North America	Immigration				
Latin America	Emigration				
Russia					
Middle East	Decline				
Africa	Increase				
Australia	Immigration				
Europe	Decline				

Summary

Population growth, climate change, and natural resource depletion are likely to produce significant social, political, and economic consequences that will differ by region or country.

The population of South Asia is expected to grow by 30-35% by mid-century and then stabilize. The Muslim population of the region will grow faster than the Hindu population and other minority religions. Total population growth will place a major burden on the region. However, economic growth is expected to be strong. Per capita GDP is expected to grow by 3-4% or more per year through 2100, although this growth will not be evenly distributed.

Under employment and economic disparity are likely to continue or increase. Cultural norms will lower the risk that these demographic changes will cause social unrest, at least in India. Urbanization is expected to accelerate through mid-century with four cities growing beyond 30 million in population. This trend will cause stress in some rural areas and could lower agricultural productivity. Some areas of these cities are currently slums. Providing housing and basic services to prevent social unrest and violence will continue to be a challenge.

Climate change poses a major threat to the region. A sea level rise of 1 meter or greater and intensification of violent storms would likely have a devastating effect on millions of people. A reduction in rainfall due to changes in the monsoons would reduce agricultural productivity and threaten food security.

The region does not have adequate natural resources to support the population. Absolute freshwater scarcity and inadequate arable land will become the norm as the population grows, especially if environmental conditions deteriorate. Coal will continue to be the predominant source of energy unless the region imports much larger amounts of oil and natural gas or builds much larger capacity for producing nuclear or renewable energy. Acquiring energy or building the means to produce it will divert investment from other sectors of the economy.

The region suffers from poor governance. India has the highest functioning government. The lack of effective government in the region, and especially the lack of accountability to the populace, increases the likelihood that risks will not be well managed and that needed investments will not be made to avoid adverse political, social, and economic consequences.

The population of East Asia is expected to decline and age significantly by 2100. These demographic changes will be a major economic and social challenge for the region unless people remain in the workforce after age 65, the productivity of younger workers increases dramatically, and/or people remain much healthier later in life. Immigration is unlikely to increase because of social, language, and political barriers despite the high level of economic opportunity in the region. Emigration will continue but should slow due to economic growth within the region. Economic growth is expected to be strong driven by the industrial, technology, and financial sectors. Income disparity should widen causing social changes that could be a serious challenge.

Climate change will be a major threat to the region. Japan, the Koreas, and the coast of China are vulnerable to inundation and salt-water incursion. The Gobi Desert will expand but this should not have a major impact on agricultural productivity or displace large numbers of people. Freshwater scarcity is a risk, especially in China. As diets change due to economic growth, East Asia has been importing products with a high water footprint. Reduced availability of seafood as a result of climate change would also have a major effect on diets in the region. East Asia is likely to be at high risk from violent storms that increase in severity; millions of people and large tracts of arable land could be lost.

The region does not have adequate fossil fuel reserves for the remainder of the century. Coal will continue to be the predominant source of energy. The region will need to import much larger amounts of oil and natural gas or build much larger capacity for producing nuclear or renewable energy.

China has an authoritarian highly centralized government that has proven its ability to promote economic growth and make investments for the future. Japan has highly effective government and a high level of stress tolerance. South Korea has less stress tolerance but effective government. North Korea has a highly authoritarian government with no demonstrated ability to effectively manage stress.

The population of Southeast Asia will continue to grow modestly until 2050 and then stabilize. Indonesia will remain the largest country in the region. Emigration to countries with greater economic opportunity will continue and could slow economic growth, especially in the Philippines. Urbanization, especially among young people, will continue, leaving some rural areas without an adequate workforce. GDP growth is expected to be strong from the agricultural and industrial sectors.

Climate change will pose a major risk because of sea level rise and an increase in violent storms. Loss of arable land due to poor water management and inundation of coastal regions is a risk. The region does not have significant energy reserves or nuclear power capability. The region will need to build a large renewable capability, acquire nuclear power, or import large amounts of energy to meet demand and support economic growth. An inability to provide power for economic growth is a major risk. Government effectiveness in the region is variable. None of the countries have a high level of stress tolerance but the risks for the region are moderate. Most of the problems will center on adapting to climate change and generating the energy needed for continued economic growth.

By 2100, the population of North America will grow by nearly 40% through immigration, changing the cultural demographics of the region. People of Hispanic and Asian descent will be more than 40% of the population by the end of the century. If immigrants join the labor force and are productive, GDP growth should remain 1.5-2.5% and could be higher. US debt to foreign nations will slow US GDP growth, as will persistent trade deficits. Aging of the population will be a serious economic burden unless people over the age of 65 remain employed, growth in health care costs are controlled, and/or people remain healthy longer. The economic burden of increasing debt, trade deficits, and an aging population will reduce investment in infrastructure and social programs.

Climate change will have a major impact. Coastal regions will be inundated, desert regions will expand, and the strength of violent storms will increase, affecting millions of people in coastal regions and central North America. Canada will fare much better; the climate will warm, and precipitation could increase. North America is rich in natural resources and should not experience any severe shortages through the end of the century. Because of climate change, the amount of freshwater and arable land in Canada could increase significantly. The region has effective governments and a high level of stress tolerance. These strengths combined with the relatively low risks suggest that the stress from immigration and climate change are likely to be effectively managed.

The Latin American population is expected to grow to about 800 million people by mid-century and then remain stable or decline slightly through 2100. Brazil and Mexico will remain the largest countries. The rate of emigration from Latin America will remain high from those countries with low economic growth or those countries unable to reasonably ensure the personal safety of their populace. Per capita GDP growth is expected to be good in Brazil, Mexico, Argentina, and Columbia. Other countries, especially Venezuela, may fare much less well depending on the prevailing social and political environment.

Climate change will have a major impact. Coastal regions will be inundated, desert regions will expand, and the strength of violent storms will increase, affecting millions of people living on islands in the Caribbean and in the Yucatan Peninsula. The distribution of natural resources in Latin America is very uneven. Venezuela has over 90% of the proven oil reserves and more than 75% of the proven natural gas reserves. Brazil has large oil and natural gas reserves off the coast of Rio de Janeiro, but these reserves will be expensive to bring into production. Brazil and Columbia have most of the coal reserves. These disparities could lead to conflict. Latin America has ample freshwater and arable land, although some countries, such as Mexico, will continue to experience freshwater stress. Central America and South America are using agricultural land and freshwater to grow livestock for export.

The governments of Latin America have low to very low stress tolerance. Corruption control, regulatory effectiveness and the rule of law are weaknesses. The region has not demonstrated an ability to foster effective intergovernmental cooperation so the disparity in resources within the region may be a catalyst for conflict and an inability to effectively manage future challenges.

The population of Europe is expected to decrease by 10% or more by the end of the century. In addition, Europe's population is aging. More than 25% of the population will be over the age of 65 and more than 10% of the population will be over the age of 80 by 2100. Immigration into Europe has increased dramatically in the last 10 years and is expected to continue. These demographic trends will place a large burden on European economies, especially given the financial support and social services provided by many European governments. Most immigrants will come from Africa and the Middle East and will likely be Moslem. To the extent that these immigrants enter the workforce and are productive they will help support the aging population and increase economic growth. These benefits may be offset by disruptive cultural conflict that could grow in intensity if effective assimilation of immigrants is not achieved. Per capita GDP growth is expected to be modest, in the range of 1-1.5%, reflecting the maturity of these economies and relatively low level of personal consumption and business investment.

Climate change will have an impact largely through sea level rise and inundation of coastal regions. Europe has adequate freshwater and arable land, and it is unlikely that climate change will cause a critical reduction in these natural resources. Europe will likely continue to export high water footprint agricultural products to countries experiencing water stress or lacking adequate agricultural land. Europe does not have adequate energy reserves and will need to continue to import oil and natural gas. Europe will be increasingly dependent on foreign sources of energy unless nuclear and/or renewable energy grows at a much greater rate or coal reserves are used to a greater degree. Europe is among the most effectively governed regions in the world with a high level of stress tolerance but will need to solve its energy problems and effectively manage the demographic changes that will unfold due to aging and immigration.

Russia has a declining population that is aging. More than 20% of the population is expected to be over the age of 65 by 2100. A large number of illegal transient immigrants move into and out of Russia to fill jobs. Legal immigration is expected to increase due to climate change and Russia's labor force needs over the next 50 Years. Many of these immigrants will come from Africa and the Middle East and will likely be Moslem. Effective assimilation of these immigrants will be a challenge for Russia. To the extent that these immigrants enter the workforce and are productive they will help support the aging population and increase economic growth. These benefits may be offset by cultural conflict that will be disruptive and could grow in intensity if effective assimilation of immigrants is not achieved. Per capita GDP growth is expected to be modest, in the range of 1.5-2.0%, reflecting the low level of personal consumption and business investment.

Climate change will not have a major negative impact largely due to the geography of Russia. If the tundra melts and precipitation moves to higher and warmer latitudes, Russia may actually benefit from climate change. Russia already has ample freshwater and arable land and can be expected to export products with a high freshwater footprint if the growing season is prolonged due to climate change. Russia has large reserves of fossil fuels, especially natural gas, and can be expected to continue to export energy to Europe and East Asia. Arctic melting will likely add large undiscovered fossil fuel reserves over the next 50 years. Russia has an authoritarian government focused on exploiting Russia's natural resources and strengthening its military capabilities. The country failed to assimilate democracy at the outset of the 21st century so the government is not accountable to the populace.

Fortunately, Russia is subject to low levels of stress. The country is rich in natural resources and is not likely to experience serious consequences from climate change. If the country can continue to grow economically with an aging population and can manage any significant increase in immigration, it should prosper through the end of the century.

The population of the Middle East is expected to grow by more than 75% through the end of the century. Population growth will occur primarily in Egypt, Iraq, Yemen, and Syria, regions that have recently been affected by civil or military conflict. Egypt, Turkey, Iran, and Iraq are projected to be the largest countries in 2100. Urbanization and emigration to countries with greater economic opportunity are likely to increase as the climate changes. Economic growth will vary by country and will be determined by the health of the markets for fossil fuels and the degree to which Islamic sects and regional governments are able to maintain peaceful relationships. Loss of human capital and decreased agricultural productivity, accentuated by climate change, could slow economic growth.

Climate change will pose a major risk because of sea level rise and loss of arable land. The region has large fossil fuel reserves but little nuclear or renewable energy infrastructure. The region will need to continue to increase desalination capacity to meet personal and industrial freshwater needs.

Governance in the region is authoritarian and heavily influenced by Islamic law. An inability to bridge religious differences between Islamic sects has led to civil and military conflict that has taken millions of lives and led to economic deprivation for many in the region. The lack of effective governance and the persistence of religious conflict reduce the likelihood that the region will effectively manage the stresses induced by population growth, freshwater scarcity, lack of arable land, and climate change.

The population of Africa is expected to grow by more than 300% by the end of the century, assuming climate change doesn't significantly reduce fertility rates in Sub-Saharan Africa. Population growth will occur primarily in Sub-Saharan Africa. By 2100 five of the 10 largest cities in the world will be in Africa. Urbanization will create cities that will be very difficult to govern and supply with the essentials of life. Desperation could drive some to seek a better life by any means available. Emigration to countries with greater economic opportunity will increase as the climate changes. Economic growth will vary by country and will be determined by the climate, the productivity of the labor force, and the availability of capital. Loss of human capital and decreased agricultural productivity, accentuated by climate change, will slow economic growth.

Climate change will pose a major risk because of loss of arable land to desertification and scarcity of fresh water. The region has adequate fossil fuel reserves but little nuclear or renewable energy infrastructure. Africa is the second largest continent and has a large amount of natural resources but the rate and extent of human population growth over the next 35-50 years is unprecedented in human history and will cause economic hardship, environmental degradation, political instability, and civil unrest. Urbanization will be beyond the ability of regional governments to manage. Mass migration from sub-Saharan Africa will ensue.

The population of Australia is expected to increase by over 70% by the end of the century, largely as a result of immigration from India, China, and Southeast Asia. In addition, Australia's population is aging. More than 20% of the population will be over the age of 65 and more than 5% of the population will be over the age of 80 by 2100. Immigration into Australia has increased in the last 10 years and is expected to continue to add about 200,000 to 250,000 migrants per year. To the extent these immigrants enter the workforce and are productive they will help support the aging population and increase economic growth. These benefits could be offset by cultural conflicts. Per capita GDP growth is expected to be modest, in the range of 1.3-1.5% but could be higher if exports increase and the balance of trade increases.

Climate change will have an impact through sea level rise, desertification, and an increase in severe storms. Australia has adequate freshwater and arable land, and it is unlikely that climate change will cause critical shortages of these natural resources. Australia will likely continue to export high water footprint agricultural products to countries experiencing water stress or lacking adequate agricultural land. Australia has adequate fossil fuel reserves but is heavily dependent on coal. Oil and natural gas reserves will not last until 2100. Australia has little renewable energy infrastructure and no nuclear power plants, yet the country has large uranium reserves. Australia is a large continent and is rich in natural resources. Its resources should be adequate to support the population through the end of the century as long as migration does not increase significantly. The governance of Australia is highly effective and should be able to effectively manage stress related to immigration and climate change. Australia will need to husband its resources carefully and aggressively build a nuclear or renewable energy capacity.

CHAPTER 6

The Path Forward

"A bend in the road is not the end of the road, unless you fail to make the turn."
—Helen Keller

Chapter Guide

THIS CHAPTER BUILDS ON THE knowledge gained in previous chapters and summarizes the principal goals that must be achieved in order to create a sustainable world by the end of this century. Each goal is discussed in some detail, including case studies or examples to illustrate available alternatives. Existing international or country-specific mitigation programs are reviewed. Finally, a set of specific objectives are described in order to illustrate the steps that could be taken to address population growth, climate change, and natural resource depletion. Use of economic or political incentives or disincentives is discussed, as is the role of technology development. A summary concludes the chapter.

Introduction

Over the past 40-50 years there have been many programs and initiatives developed by governments, the United Nations, and non-governmental organizations to reduce population growth, mitigate climate change, and conserve natural resources. New technologies and capabilities have been developed and deployed, international agreements have been signed and implemented, many scientific and policy forums have been held, countless books, articles, and newspaper accounts have been published, and billions of dollars have been spent. Nevertheless, success has been elusive.

The world population growth rate currently exceeds 1.0% per year and population growth in Africa, India, and the Middle East is expected to add at least 2.0 billion people to world population in the next 30 years. Each year atmospheric carbon dioxide-equivalent concentrations increase 3-4 ppm, world energy consumption increases by 8-16 EJ, and depletion or contamination of non-renewable natural resources continues unabated. By the end of this century, if these trends persist, the world population will exceed 11 billion, the global mean surface temperature relative to the early industrial period will increase greater than 3.0°C (5.4°F), Earth's economically recoverable oil and natural gas reserves will be largely depleted, and fresh water and arable land will become scarce in several regions of the world.

Despite efforts to change course, despite the significant accomplishments of several governments, the UN, and non-governmental organizations, despite the hard work of many dedicated scientists, professionals, and field workers, and despite the sizable investments that have been made, the storm clouds are still gathering on the horizon.[120] At the end of this century, if we fail to take "the bend in the road" now, mankind could face an existential threat in the 22nd century. A lot has been done but we have to do more, much more, if we want to avert overpopulation, global warming, severe environmental degradation, and scarcity of essential natural resources.

Of course, there is another view. World population growth rates are decreasing and could be less than 0.1% per year by the end of the 21st century. The Earth can sustain a stable human population of at least 10 billion, especially if new technology continues to improve agricultural productivity. Market forces will drive the development of new transformative technologies that will substitute for reduced availability of non-renewable natural resources. Current international agreements will drive reduced carbon emissions. Natural climatic cycles will limit global warming to less than 1.5°C this century. The energy industry will develop new technologies, adapt to market needs, and provide adequate energy to meet future demand. International trade will get food and natural resources where they are needed. Mankind will adapt, as it always has, and there is no reason to adopt very costly, distortionary policies that could disrupt global GDP growth and the current world political order, especially for the US, China, and India.

If the first scenario is more likely, as current evidence would suggest, a rational approach is to develop strategies to reduce population growth and carbon emissions, to adapt to climate change, and to conserve natural resources. Fortunately, effective methods to achieve each of these goals already exist. Furthermore, new technologies and new methods are being developed, although the timing, effectiveness, and cost of these new technologies are difficult to determine until they have been effectively scaled up.

The challenge is to develop, finance, and implement strategies that will reduce the threat posed by the interaction of population growth, climate change, and natural resource depletion. This means integrating population control and climate change programs and incorporating conservation of natural resources into existing programs and strategies. These strategies should leverage the scientific, governmental, and humanitarian efforts

---

120 D. Tong, Q. Zhang, Y. Zheng, et al., "Committed Emissions from Existing Energy Infrastructure Jeopardize 1.5 °C Climate Target," *Nature* 572 (2019): 373–377, https://doi.org/10.1038/s41586-019-1364-3.

already in place. Programs and strategies that address one of these threats but not the others may achieve narrow program goals but will most likely fail to contribute meaningfully to building a sustainable world in the 22nd century.

It seems abundantly clear that mankind must do more than it has during the last 30 years if we are to to build a sustainable world in the second half of the 21ˢᵗ century. In order to achieve this mission, the following goals must be met:

- First, we must develop plans to accelerate the current decline in the population growth rate in those countries where the growth rate exceeeds the replacement rate. The long-term goal should be to eliminate positive population momentum and maintain the world population at 9±0.2 billion in 2050 and 8.5 billion or less in 2100.
- Second, we must eliminate the global net increase in atmospheric carbon emissions as soon as possible.
- Third, we must increase investment in new technology and/or improve existing technology that will:
  - Improve energy efficiency
  - Reduce greenhouse gas emissions or atmospheric carbon concentrations
  - Develop alternatives for producing petrochemicals
  - Produce fresh water and protect existing ground water reserves
  - Improve crop resistance and increase crop productivity
  - Promote adaptation to global warming.
- Finally, we must develop practical and effective methods to finance and implement mitigation programs by nation or region.

Development and implementation of these strategies will require complex and politically difficult international agreements fully supported by the developed world and led by China, India, and the United States. Sacrifices and rationing will be necessary, and they will be especially difficult in the developing world. The international community will need to find ways of increasing cooperation, holding governments and governmental leaders accountable, and enforcing compliance through judicious use of political and economic rewards and penalties.

Population Control

Goal 1: We must actively accelerate the current decline in the population growth rate in those countries where the growth rate exceeds the replacement rate so as to eliminate positive population momentum and maintain the world population at 9±0.2 billion in 2050.

The UN population division has constructed global and country-specific population projections based on different assumptions about current and future total fertility, mortality, and migration rates. The UN medium variant estimate, which assumes the current rates of decline in total fertility and all causes of mortality for each country persist, is a world population of 9.7±0.6 billion in 2050 and 11.0±1.6 billion on 2100. In 2008 the world population growth rate was 1.24% and in 2018 it was 1.11%, which is a decrease of 0.013% per year. If the world population growth rate continues to fall 0.013% per year through the end of the century, the world population will be about 10 billion in 2050 and about 12 billion in 2100, which is consistent with the UN medium variant projections.

There is a high level of positive population momentum in the developing world (Figure 6.1) so even significant reductions in the total fertility rate will not result in negative population growth for many years. The developed world has negative or no population momentum, which should lead to a slight decrease in population size, if migration from the developing world to the developed world is not accounted for. The UN momentum variant assumes the total fertility rate in every country drops to replacement levels immediately and trends in mortality rates remain unchanged. The momentum variant projects a global population of 9.1 billion in 2050 and 9.3 billion in 2100.

Figure 6.1. Population pyramid for the less developed and more developed regions

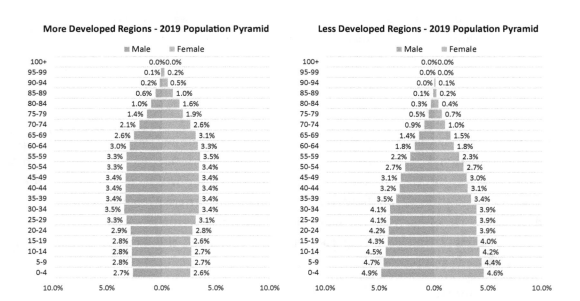

The UN low-variant projection holds assumptions about life expectancy and migration constant and varies the assumption about changes in total fertility rates from 2019 to 2100. The low-variant scenario assumes the total fertility rate is 0.5 children less than

the medium-variant scenario for each region in each year. The low-variant scenario projects world population of 8.9 billion in 2050 and 7.3 billion in 2100. Although the underlying assumptions of the low-variant scenario would be very difficult to achieve, these projections, combined with the population control experience in East Asia, suggest it may be possible to achieve a world population of 9 billion in 2050 and 8.5-9.0 billion in 2100.

There are a number of social and economic factors that can reduce population growth. These factors include urbanization, growth in per capita GDP, and changing cultural norms. Factors that can increase population growth are a lack of reproductive healthcare for women, certain cultural beliefs, poverty, lack of secondary education, and dependence on children for support in later life. Cultural beliefs include certain entrenched religious tenets or traditional practices. For example, some religions do not approve the use of birth control methods and some religious sects do not condone educating women, employing women, or promoting sexual equality. Relying on social and economic forces, such as urbanization, to produce sustained reductions in population growth in the developing world is a risky strategy since the relative impact of these various factors may vary greatly from region to region.

A number of effective direct population control strategies exist. These strategies can be broadly classified as those that interfere with reproductive biology, those that rely on changing human behavior, and those that are directly or indirectly coercive through changes in public policy. Strategies that interfere with reproductive biology include use of contraceptives, voluntary abortion, voluntary sterilization, and promotion of breast-feeding. Breast-feeding is a means of delaying the return of fertility after delivery. Contraceptives are best used in combination with behavioral strategies and are generally ineffective in reducing population growth rates when used alone.

There are three basic types of contraceptives: those that are based on administering hormones to prevent ovulation, those that present a barrier to fertilization, and those that prevent intra-uterine implantation of an embryo. Hormonal contraceptives use estrogen and progesterone in varying doses and schedules to prevent ovulation. These hormones can be administered as a pill or as a depot formulation that can prevent ovulation for many months or years. Depot formulations include injections, subcutaneous implants, vaginal rings, intra-uterine systems, and cutaneous patches. Contraceptive barriers include both male and female condoms, diaphragms and cervical caps. Finally, implantation of an embryo can be prevented by use of various intra-uterine devices such as coils or loops. Use of abortion and sterilization procedures must be regulated to ensure they are voluntary and do not violate accepted cultural or religious norms. Most of these methods are reversible and can be effective in reducing the total fertility rate if used in combination with behavioral methods.

Behavioral and social strategies can be effective in reducing the total fertility and population growth rates. Improving maternal and child health care is especially important. Lower infant and child mortality encourages parents to opt for smaller families. If parents think some of their children will die, they want to have several to make sure that some survive. However, if they expect their children to live, they want to invest in their children's future to maximize their opportunities in life. In general, healthier families are smaller families.

Healthcare need not be provided by highly trained and expensive medical professionals; paramedics can provide competent health care, especially in rural or remote areas. Family planning services are also effective, especially specific governmental programs that employ local field workers. For example, using local midwives in family planning is helpful. Educating men and women through at least secondary school and including sex education in the curriculum for all students is important. Promotion of gender equality has also been shown to be effective and should include actively assisting women who want to enter the workforce. Efforts to reduce or abolish gender bias from law, business, and culture can reduce fertility rates.

Another effective strategy is promotion of late marriage, such as delaying marriage until after age 25. By delaying marriage, the number of childbearing years for each woman is reduced. Promoting abstinence is an ineffective strategy, especially when used alone. In general, effective population control programs rely heavily on one or more of these behavioral strategies, although religious or cultural norms can pose a significant barrier to their adoption. Tailoring family planning to the specific, regional cultural and economic environment and employing local workers is essential for success.

Some countries have found it necessary to adopt coercive public policy to achieve a desired reduction in the population growth rate. These coercive strategies generally involve financial incentives or disincentives through taxes, subsidies, rebates, or credits. In some cases, certain essential services, such as healthcare, are either provided or denied depending on the number of children in a household. These coercive strategies promote one or two child families and discourage having three or more children. In general, these coercive policies have been effective when adopted by central governments, but they raise ethical questions that may need to be considered.

The major national or international efforts to control population growth are:

1. *UN Population Action Plan:* The UN Population Action Plan was adopted in 1974 and re-affirmed in 1994. The Plan establishes 15 principles that define the values that guide program development. The Plan defines human reproductive rights, respects cultural and religious beliefs, recognizes the sovereignty of individual countries, and addresses population growth in the context of sustainable

economic development. The Plan calls for governments to include population control in economic development plans and programs, reduce poverty, implement educational programs, promote gender equality and empower women, provide family planning services and technology (e.g., contraceptives), and other actions designed to reduce the country-specific population growth rates.

2. *Individual National Governmental Programs:* Many sovereign countries, especially in the developing world, have population control programs, some of which are voluntary and others of which are coercive.

3. *Non-Governmental Organizations:* There are more than 80 major global non-governmental organizations that focus to varying degrees on population control and related issues around the world. These NGO's receive private and governmental financial support and conduct programs ranging from demographic research to field service, often in conjunction with the UN.

In April of 2019, the UN issued a press release summarizing the conclusions of the 52nd session of the United Nations Commission on Population and Development. (1) The UN concluded that "…over the past 25 years or more significant progress has been made by the collective effort to promote population control. Examples include greater access to sexual and reproductive health care, reduced child and maternal mortality, increased life expectancy, and advances in gender equality and women's empowerment. Between 1994 and 2019, the total fertility rate fell from 2.9 to 2.5 births per woman. However, fertility levels remain high in sub-Saharan Africa (4.8), Central and Southern Asia (2.4), Oceania excluding Australia and New Zealand (3.4), and Western Asia and Northern Africa (2.9)… Use of modern methods of family planning has risen markedly. Globally, among married or in-union women of reproductive age who express a need for family planning, the proportion for whom such need is satisfied with modern methods of contraception, increased from 72 per cent in 1994 to 78 per cent in 2019. Nonetheless, in 44 countries, mostly in sub-Saharan Africa and Oceania, less than half of women's demand for family planning is being met by use of modern methods. Globally, the birth rate for those aged 15 to 19 declined from 65 births per 1,000 women around 1994 to 44 per 1,000 currently. The highest levels of adolescent childbearing are found in sub-Saharan Africa (101 births per 1,000 adolescent women) and in Latin America and the Caribbean (61 births per 1,000 adolescent women)." The UN concluded "we must prepare for a world with a population that is larger, older, more mobile and more urbanized than ever before."

Case Studies

South Korea emerged from the Korean War (1950-1953) with a population of about 25 million people. In 1955, over 40% of the population was less than 15 years of age, the population growth rate was 2.9%, and the total fertility rate was 6.1 live births per woman of childbearing potential. In 1962 the government began a nationwide family planning program because population growth was believed to be undermining economic growth. Both public and private agencies were engaged in developing and implementing local family planning programs. In 1973 a Maternal and Child Health Law legalized abortion under certain circumstances. In 1983 the government began suspending medical insurance benefits for maternal care for pregnant women with three or more children. It also denied tax deductions for education expenses to parents with two or more children. In 1983 the population growth rate was 1.5% and the total fertility rate was 2.9 births per woman. In 1985, just two years later, the population growth rate was 1.0% and the total fertility rate was 1.66. In the late 1980s, other programs were instituted, including distribution of free birth control devices and information, classes for women on family planning methods, and the granting of special subsidies and privileges (such as low-interest housing loans) to parents who agreed to undergo sterilization.

Other factors that contributed to a slowdown in population growth included urbanization, later marriage ages for both men and women, higher education levels for women, a greater number of women in the labor force, and better health standards. Korea's economy has grown dramatically since 1980; per capita GDP increased from $1,700 in 1980 to $12,000 in 2000 and was >$31,000 in 2018. The compound annual growth rate for per capita GDP during this period is a remarkable 7.9% per year.

In 2000 the population of South Korea was 46 million, the population growth rate was 0.8% and the total fertility rate was 1.47. In early 2019, the total fertility rate decreased to 0.98 live births per woman of childbearing potential and the population growth rate was less than 0.4%. Today Korea has negative population momentum in addition to a low total fertility rate. Only 13.8% of the population is under 15 years of age. By 2023 the population of South Korea is expected to stabilize at about 53 million and then will begin decreasing. South Korea is now aging faster than any developed country in the world. The South Korean government has called attention to the aging population and its implications for economic growth and has discussed new governmental policies that will encourage women to have more children. South Korea has not actively sought to attract immigrants to increase the number of young people in the workforce, possibly because of the cultural consequences of such a policy. At the same time, the development of advanced robotics and artificial intelligence is likely to reduce labor demand in the manufacturing sector. The net economic impact of these population dynamics is yet to be determined.

Countries from other regions of the world have also successfully implemented policies that have reduced population growth rates. Bangladesh and Botswana are examples. Pakistan is an example of a country that has not succeeded and continues to have high total fertility and population growth rates. Table 6.1 compares the population growth rate and total fertility rate for South Korea, Bangladesh, Botswana, and Pakistan in 1960,

1990, and 2017. In all of these countries, per capita GDP has increased since 1960. All four countries had per capita GDP below $200 in 1960. South Korea's economy has grown dramatically, and per capita GDP was ~$31,300 in 2017. South Korea is now an industrialized country with a mature economy. Botswana's economy benefitted from the growth of the diamond industry in the 1980s. By 1990 per capita GDP was ~$3,000 and in 2017 was $8,300. Bangladesh and Pakistan have fared less well. Per capita GDP in both countries was less than $1700 in 2017.

Table 6.1: Population growth rate (PGR) and total fertility rate (TFR) for South Korea, Bangladesh, Botswana, and Pakistan in 1960, 1990, and 2017

| Country | Year | | | | | |
| | 1960 | | 1990 | | 2017 | |
	PGR	TFR	PGR	TFR	PGR	TFR
South Korea	2.9	6.1	1.0	1.6	0.4	1.2
Bangladesh	2.8	6.7	2.5	4.5	0.9	2.1
Botswana	2.3	6.6	2.8	4.5	1.8	2.7
Pakistan	2.3	6.6	2.9	6.0	2.0	3.5

Pakistan and Bangladesh are both Muslim countries but have different official languages, Urdu in Pakistan and Bengali in Bangladesh. The two countries were separated in 1972 and over the last 40 years have developed differently. For example, the literacy rate in 2018 for adult females was 71% in Bangladesh and 46% in Pakistan. The current GDP per capita for these two countries is similar. Pakistan is much larger than Bangladesh (310 sq. mi. vs 57 sq. mi.) but has a much more varied geography. The consequences of current population growth rates between Bangladesh and Pakistan are substantial. By 2050 the population of Bangladesh should stabilize at ~200 million while the population of Pakistan will continue to grow and could surpass 350 million. By 2100, if current trends persist, the population of Pakistan could be more than double that of Bangladesh. The social, economic, and political consequences of this differential population growth are likely to be significant.

There are some common themes that emerge from the experiences of South Korea, Bangladesh, Botswana, and other population control programs:

- Urbanization, economic development, and changing cultural norms are associated with decreased total fertility rates and slower population growth but the magnitude of their effect varies by region.
- Population control programs can be effective when they are promoted and managed by stable governments, when experienced non-governmental organizations

are actively involved, and when coercive governmental policies are judiciously employed.

- Religious and cultural norms can be a major barrier to reducing population growth.
- Population momentum has a major impact but can be overcome by sustained efforts to reduce population growth, although it can take 30 years or more to achieve meaningful changes in population momentum.
- While population control programs have been effective in East Asia, it is not clear if similar methods will work in Sub-Saharan Africa, South Asia, or the Middle East.

The Path Forward

Reducing global population growth will reduce global carbon emissions, environmental degradation, and the rate of natural resource depletion, if we improve our use of technology and avoid wasteful policies and practices. Achieving a world population of less than 8.5 billion in 2100 will be very difficult, requiring setting targets and developing strategies to achieve intermediate goals. Table 6.2 summarizes a set of hypothetical targets that are aggressive but within the range of plausible achievement based on UN projections.

Table 6.2: Population targets for 2050 and 2100

Population	2019 Population	2050 Current Population Estimate	2050 Proposed Population Target	2100 Proposed Population Target
Developed world	1.3	1.3	≤1.2	≤1.2
Developing world	6.5	8.4	≤7.8	≤7.3
Total	7.8	9.7	≤9.0	≤8.5

In 2018 the total fertility rate in the developing world was 2.5 live births per woman of childbearing potential (Range: 1.4 to 7.2); 50 of 137 developing countries had a total fertility rate greater than 3.5. There are 10 African and 4 Middle Eastern countries that will account for more than 90% of the projected population growth through 2100. In order to meet the population goal in 2100 the total fertility rate in these 14 countries must be lowered substantially. The 10 African countries are listed in Table 6.3. Most of these countries are in Central or West Africa.

Table 6.3: African countries with high population growth and total fertility rates

Country	GDP ($ Billion)	2018 Population (Million)	2100 Population (Million)	Population Growth Rate (Percent/ year)	2019 Total Fertility Rate (Children/ woman)
Nigeria	397	196	752	2.6	5.4
Niger	9	22	209	3.8	7.0
Mali	17	19	93	3.0	5.9
Chad	11	16	69	3.0	5.8
Angola	106	28	139	3.3	5.6
Republic of Congo	47	84	389	2.6	6.0
Uganda	27	43	203	3.7	5.0
Ethiopia	84	109	243	2.6	4.3
Kenya	88	51	157	2.3	3.5
Sudan	41	42	127	2.4	4.4
Total	827	610	2,381	2.6	>5.0

Africa is a diverse continent that has undergone major political, economic and social changes over the last 50 years. Mortality rates and especially infant mortality rates have decreased significantly in most countries, resulting in a substantial increase in life expectancy. Becoming healthier and living longer was associated with substantial reductions in total fertility rates in East Africa but not in Central or West Africa. (Table 6.4) Unfortunately, there has not been a corresponding increase in per capita GDP and secondary education rates in any of these regions.

Table 6.4: Comparison of TFR in 1984 and 2017 in East Africa and West/Central Africa

Country	1984 TFR	2017 TFR	Percent reduction from 1984
East Africa			
Ethiopia	7.44	4.08	45.2
Kenya	7.04	3.79	46.2
Rwanda	8.28	3.81	54.0
West and Central Africa			
Nigeria	6.73	5.46	18.9
Niger	7.88	7.18	8.9
Angola	7.39	5.62	23.9
Mali	7.15	5.97	16.5

Sub-Saharan Africa is a patriarchal society that traditionally places a high social premium on women having many children. Villages and extended families provide care for infants and children when parents are engaged in activities outside of the home. Marriage rates for women are very high and early marriage is nearly universal. Polygamy is acceptable in some cultures, exacerbating conditions that promote high total fertility rates. In the past, high infant mortality rates and social norms that discouraged post-natal intercourse and promoted breast-feeding acted to limit population growth in Sub-Saharan Africa.

In the last few decades however, infant mortality rates have dropped as a result of improved nutrition and sanitation and the availability of child healthcare. Social norms have also changed, resulting in less reliance on breast-feeding. Collectively, these social forces have increased population growth rates. Those African countries that have been able to reduce the total fertility rate, such as Rwanda, Kenya, and Ethiopia, have done so through vigorous state-led programs that promote monogamy and provide family planning programs, including distribution of birth control methods and by educating both men and women through secondary school. Use of coercive methods or enacting laws to prevent childhood marriage and promote gender equality have not been widely adopted in Africa.

Involvement of religious leaders in promoting family planning and use of contraceptives is essential to reducing total fertility rates in Sub-Saharan Africa. For example, the total fertility rate in the Moslem northern region of Nigeria is twice as high as the Christian south. Religious beliefs are a source of cultural differences. Christianity is the largest religious group in the world with about 2.4 billion followers, representing 31% of the world's population. Islam is second with about 1.9 billion adherents or 24% of the world's population. Agnostics and Hindu's follow with about 1.15 billion each. This distribution is changing because the population growth rate among Muslims is 1.5% per year, which is almost twice the population growth rate for all other religions and agnostics combined. The high population growth rate among Muslims is due to a high fertility rate, estimated to be about 2.9 live births per woman of childbearing potential compared with 2.1 births per woman of childbearing potential in non-Muslims, and strong positive population momentum in Muslim countries. There are four Moslem countries with high total fertility and population growth rates: Egypt, Iraq, Pakistan, and Algeria. (Table 6.5)

Table 6.5: Middle Eastern countries with high population growth and total fertility rates

Country	GDP ($ Billion)	2018 Population (Million)	2100 Population (Million)	Population Growth Rate (Percent/ year)	2019 Total Fertility Rate (Children/ woman)
Egypt	251	98	201	2.0	3.3
Iraq	226	38	164	2.3	3.7
Pakistan	313	212	364	2.1	3.4
Algeria	181	42	61	2.0	3.0
Total	961	390	790	2.2	3.5

Several Moslem countries have population growth rates greater than 1% but less than 2.0%. (Table 6.6)

Table 6.6. Middle Eastern countries with population growth rates less than 2.0%

Country	GDP ($ Billion)	2018 Population (Million)	2100 Population (Million)	Population Growth Rate (Percent/ year)	2019 Total Fertility Rate (Children/ woman)
Saudi Arabia	782	34	48	1.8	2.5
Iran	454	82	70	1.4	1.6
Turkey	767	82	88	1.5	2.0
Indonesia	1,042	269	314	1.1	2.3

Case Studies

In 1975 Iran had a population of 33 million. In the decades prior to 1975 Iran had a population growth rate of 2.5-2.7%, the total fertility rate was 6.2-6.9 births per woman of childbearing potential, and the infant mortality rate was greater than 120 deaths per 1000 live births. In the late 1960s the Iranian government created a Family Planning Division in the Ministry of Health and family planning clinics were established throughout the country. Despite these efforts, family planning services and contraceptives were not well accepted as is evidenced by the persistently high total fertility rate. The infant mortality rate began to decline in the 1970s and was less than 100 deaths per 1000 live births in 1976 but the total fertility rate remained high at about 6.5 births per woman.

The Islamic Revolution occurred in 1979. The new government initially encouraged married couples to have more children, dismantled the Family Planning Division and the family planning clinics, and passed laws to promote population growth. These laws included legalization of polygamy and lowering of the legal age for marriage to 9 years. Other pro-population growth policies included financial incentives to newlyweds and inflating the cost of birth control methods. The population growth rate continued to increase after the revolution and reached 4.1% in 1984. From 1968 to 1988 Iran's population doubled from 27 million to 55 million.

In the late 1980s the leadership of the Iranian government changed. Rapid population growth was viewed as a serious threat and the government's policies changed dramatically. The Ayatollah declared Islam favored having only two children and various contraceptive methods were made widely available. In 1993, the Parliament withdrew the benefits to government workers who had a third child. By 1995, the population growth rate had dropped precipitously to less than 1.5%, the total fertility rate decreased to 3.2 births per woman, and infant mortality was 35 deaths per 1000 live births. Ten years later, the government began to reverse its policies in response to the prospect of negative population growth. On July 25, 2012 Supreme Leader Khamenei stated that Iran's contraceptive policy made sense 20 years ago, "but its continuation in later years was wrong ... Scientific and expert studies show that we will face population aging and reduction (in population) if the birth-control policy continues." Subsequently, budgets for family planning have been cut, paternity and maternity leaves have been decreased, and some forms of contraception have now been made illegal.

In 1975 Pakistan had a population of 67 million, a population growth rate of 2.5%, a total fertility rate of 6.5 births per woman of childbearing potential, and an infant mortality rate of 140 deaths per 1000 live births. In the 1950s and 60s early family planning efforts were organized under a military government but were largely ineffective. During the next 50 years there were several abrupt changes in government resulting in dramatic changes in public policy. Some governments promoted family planning, and some were aligned with conservative Islamic sects and discouraged family planning. Extremely conservative Islamic beliefs predominate in many parts of Pakistan. This volatility in governmental leadership and public policy resulted in low levels of contraceptive use, ineffective and inconsistent family planning, poor maternal and child healthcare, and ineffective educational policies.

The consequence of political upheaval, traditional and religious beliefs, and neglect has been a continued rapid increase in Pakistan's population. In 2018, the infant mortality rate was still 61 deaths per 1000 live births and the total fertility rate was 3.4 children per woman of childbearing potential. Pakistan has a high level of population momentum. (Figure 6.2) In 2018, the population was 212 million and the population growth rate was 2.1%. At this growth rate Pakistan is adding nearly 4 million people each year and is projected to have a population of 360-380 million by 2100.

The experience in Iran and Pakistan demonstrates the importance of a stable government committed to controlling population growth. It also illustrates the value of programs that emphasize behavioral as well as reproductive strategies and the importance of educating women. Finally, it demonstrates that religious leaders in Islamic countries can play a very important role in changing cultural attitudes and reproductive behavior.

India

India will soon be the most populous country in the world. It is large and geographically, culturally, and politically diverse. For example, people in neighboring states may not speak the same language because there are at least 24 distinct languages spoken. In 2018 the population reached 1.353 billion and was growing at about 1.037% per year. At this rate, India adds about 13-14 million people per year. In the same year, the total fertility rate was 2.3 live births per woman of childbearing age, infant mortality was 32 deaths per 1000 live births, and per capita GDP was about $2,100. India has positive population momentum that will cause its population to continue to increase until about 2050-2070 when its population may stabilize at 1.7-1.8 billion.

Figure 6.2: Population pyramid for Pakistan

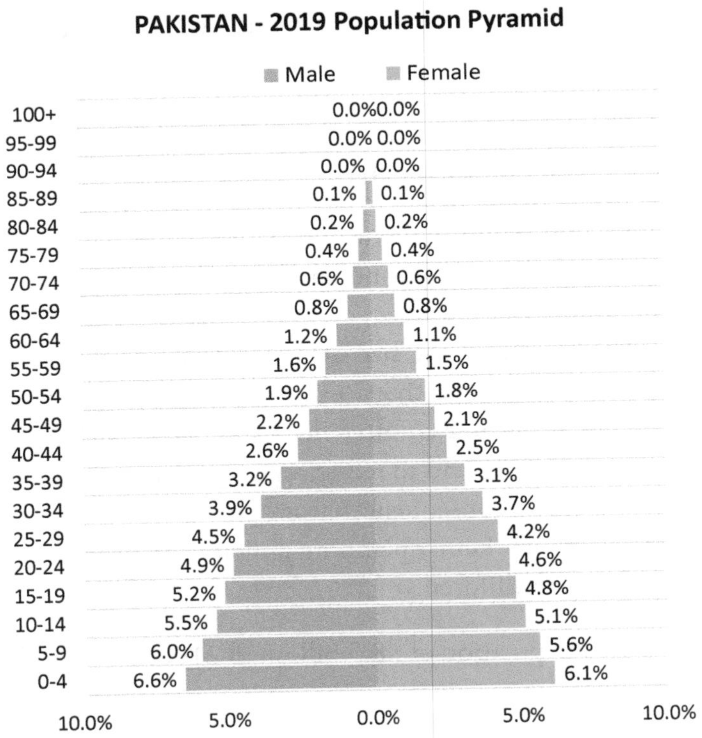

PAKISTAN - 2019 Population Pyramid

■ Male　　■ Female

Age	Male	Female
100+	0.0%	0.0%
95-99	0.0%	0.0%
90-94	0.0%	0.0%
85-89	0.1%	0.1%
80-84	0.2%	0.2%
75-79	0.4%	0.4%
70-74	0.6%	0.6%
65-69	0.8%	0.8%
60-64	1.2%	1.1%
55-59	1.6%	1.5%
50-54	1.9%	1.8%
45-49	2.2%	2.1%
40-44	2.6%	2.5%
35-39	3.2%	3.1%
30-34	3.9%	3.7%
25-29	4.5%	4.2%
20-24	4.9%	4.6%
15-19	5.2%	4.8%
10-14	5.5%	5.1%
5-9	6.0%	5.6%
0-4	6.6%	6.1%

10.0%　　5.0%　　0.0%　　5.0%　　10.0%

Religious diversity also exists. In 2018 Hindus made up about 80% of the population and Moslems about 15%. The total fertility rate (TFR) among Muslims in the region is about twice the rate for Hindus and population momentum is greater for Muslims. In 2016 the total fertility rate for Muslims was 2.6 and for Hindus was 2.1. Both rates have decreased by about 20% in the last decade. Despite a decreasing total fertility rate for both populations, population momentum will result in a Moslem population in India in 2050 of 350-375 million, up from 200 million in 2016. Should this projection materialize, Moslems will represent 20-22% of the population by mid-century and will produce 150-175 million of the additional 350 million people living in India.

If India is to achieve lower population growth and nearly 50% of the projected population growth will come from the Moslem community, the total fertility rate in the Moslem community will need to be reduced. However, many Moslems view this assertion as a Hindu effort to suppress their community. This religious antipathy makes it imperative that population control programs are equitable and not unjustly coercive. Given the challenges that India faces, the Indian government is considering legislation that would offer incentives to families for having no more than two children.

Population Decline

There are several countries that are experiencing population growth of less than 0.5% or population decline, including Japan, Russia, Germany, South Korea, and several countries in Eastern Europe. In countries that are experiencing population decline, the death rate is higher than the birth rate and immigration does not make up the difference. In other countries, the death rate is higher than the birth rate, but the population remains stable or grows slightly due to immigration. In other cases, the birth rate is higher than the death rate, but the population remains stable or declines due to emigration.

Low population growth or a declining population in a country can be associated with an increase in per capita GDP, reduced environmental degradation, and reduced depletion of natural resources. Potential adverse consequences of a declining birth rate include aging of the population and reduction in the size of the labor force. Aging of the labor force can be associated with reduced productivity, innovation, and investment in research and development. If the population of a country decreases, demand for goods and services could decrease as well and possibly lead to recession or economic stagnation. The prospect of adverse economic and social consequences from population decline has caused some developed countries to alter their family planning policies and promote an optimal family size that includes two or more children.

In order to achieve a world population of ≤8.5 billion in 2100, there are several conditions that must be met:

1. The total fertility rate in the developing world must be reduced to replacement levels or below as soon as possible.

    o The population replacement rate in the developing world varies depending on mortality rates but is typically 2.5 to 3.3%. The total fertility rate need not be reduced to 2.1 or less unless there has been a corresponding decrease in the replacement rate.

o Positive population momentum in the developing world must be reversed.
o It will take 20 to 30 years to achieve this goal.

2. The developed world must not reverse or delay its demographic transition to a stable or decreasing population size in order to promote economic growth.

In order to mitigate the adverse economic consequences of population decline, the developed world should enhance immigration policies to attract workers from the developing world, where population growth may be producing economic decline. In order for a government to maintain domestic tranquility and avoid cultural upheaval, it must maintain control of its borders and have the ability to manage immigration to meet the needs of society. Unskilled immigrants can be an asset if they are trained to fill needed jobs when they relocate. Immigrants who do not enter the labor force are an economic burden or worse. Programs for training immigrants and promoting language and cultural assimilation are very important. Those developed countries that implement effective immigration policies will have a competitive advantage.

Climate Change

Goal 2: We must eliminate the global net increase in atmospheric carbon emissions.

If we assume that global insolation at the top of the atmosphere remains stable for the foreseeable future, that a reduction in the earth's albedo from loss of polar and glacial ice is offset by an increase in albedo due to increased cloud cover, and that the heat capacity of the earth's surface doesn't change significantly, then changes in global climate will be driven by changes in the greenhouse effect.

From 1880-1910 the atmospheric concentration of CO_2-equivalents was about 300 ppm and the global mean surface air temperature (GMSAT) was 13.74 ± 0.02°C. By the end of 2020, the atmospheric concentration of CO2-equivalents had risen to 500 ppm and the GMSAT was 14.94°C, having risen 1.2°C since the early industrial period. Over the last decade the atmospheric concentration of CO_2-equivalents has increased by 3-4 ppm per year[121] and the GMSAT has increased approximately 0.04°C per year, despite the Paris climate accords and more than 30 years of efforts to reduce carbon emissions.

Based on these trends, the atmospheric concentration of CO_2-equivalents in the 2050s will likely be 590-620 ppm and the global mean surface air temperature will likely increase from 14.85°C (58.7°F) in 2019[122] to 16.3°C to 16.9°C (61.3°F to 62.4°F) by the 2050s,

---

121 Retrieved from https://www.esrl.noaa.gov/gmd/aggi/aggi.html

122 NOAA National Centers for Environmental Information, State of the Climate: Global Climate Report for 2019, published online January 2020, retrieved on January 7, 2021 from https://www.ncdc.noaa.gov/sotc/global/201913/supplemental/page-1.

a rise of 1.45°C to 2.05°C relative to 2019 and 2.56°C to 3.1°C relative to the early industrial period. The changes in global mean surface temperature will be greater over land than over oceans, will be greater in northern latitudes, and will persist for decades.

Beyond 2050, the rise in the global mean surface air temperature will be largely determined by the effectiveness of international efforts to reduce net atmospheric carbon emissions. If these efforts are successful and the atmospheric concentration of CO_2-equivalents in 2100 is 700-750 ppm, then the global mean surface air temperature will likely stabilize at 16.5°C to 17.3°C, a rise of about 3.2°C from the early industrial period. If these efforts are unsuccessful and the atmospheric concentration of CO_2-equivalents in 2100 is 1100-1200 ppm, the global mean surface temperature could reach 18.5°C or higher.

Rises in the GMSAT that are greater than 3°C will have serious or possibly catastrophic environmental consequences. Based on our current understanding, sea levels will rise a meter or more and inundate coastal regions, precipitation patterns will change resulting in increased precipitation in equatorial and higher latitudes and desertification in mid-latitudes, cyclonic storms will be more violent driven by increased ocean temperatures, ocean and atmospheric currents will change, flow in many major rivers will decrease, and agricultural zones will shift to higher latitudes. Many species of flora and fauna will become extinct. The Earth will be a very different environment. It is doubtful such a global environment would be able to sustain a human population of 10 billion. The global economic, social, health, and political consequences would likely be profound.

In order to eliminate the global net increase in atmospheric carbon emissions, three broad objectives must be met.

1. First, greenhouse gas emissions must be reduced by:

    a. decreasing the use of coal and oil and increasing the use of natural gas,
    b. increasing the use of wind, geothermal, and solar power,.
    c. increasing the use of nuclear power,
    d. improving agricultural and industrial practices to reduce methane and nitrous oxide emissions,
    e. increasing use of electric vehicles for ground transportation and developing new low-net-carbon fuels for air and marine transportation,
    f. increasing energy efficiency, and
    g. improving existing technology or developing new technology for low-net-carbon energy production.

2. Second, carbon uptake through photosynthetic production of biomass must be increased.

    h. Deforestation, net of reforestation, must be stopped.

   i.   New technology for using carbon dioxide to grow biomass must be scaled up and deployed.

3.   Third, new technology for carbon capture and storage or utilization must be scaled up and deployed.

All of these objectives need to be pursued. Addressing only 1 or 2 of these objectives will not achieve the broad goal. Installing solar panels and windmills and buying electric vehicles is good, but the impact of these actions is very unlikely to be sufficient to achieve the broader goal. However, if these objectives can be met by 2050, the greenhouse effect can be stabilized and much of the additional global warming and environmental degradation expected after 2050 can be averted.

Current Climate Change Programs

In 1988 the World Meteorological Organization created the Intergovernmental Panel on Climate Change (IPCC) with the support of the United Nations. In 1990 the IPCC issued its first Assessment Report, which summarized the relevant climate science and concluded that the Earth was warming mainly due to anthropogenic carbon emissions. In 1992 the UN convened a conference on the environment in Rio de Janeiro that developed the UN Framework Convention on Climate Change (UNFCCC). In 1995 the parties to the UNFCCC met in Berlin and outlined specific emission targets. In December 1997 the UNFCCC parties met in Kyoto Japan and adopted a protocol to address climate change on an international basis.

The Kyoto protocol set internationally binding emission targets that were heavily weighted toward OECD countries. The protocol was based on the principle of "common but differentiated responsibilities." Detailed rules for implementation were adopted in 2001 in Marrakesh (the Marrakesh Accords) and the first commitment period started in 2008 and ended in 2012. Eighty-three countries were signatories and 192 were parties to the agreement. The United States withdrew from the Protocol in 2001.

The protocol mandated that 37 industrialized nations plus the European Community cut their greenhouse gas emissions by 5% below 1990 levels between 2008 and 2012. Developing nations were asked to voluntarily comply. More than 100 developing countries, including China and India, were exempted from the treaty. The treaty also established an international trading system, which allowed countries to earn credits toward their emission target by investing in emission cleanups outside their own country. By 2012 most participating countries had not reduced their GHG emissions and global carbon dioxide emissions increased from 32.2 Gt in 2008 to 35.5 Gt in 2012, an increase of 10%.

The most recent international effort is the 2015 Paris Agreement. Under the auspices of the UN, this agreement builds on the principles established in Kyoto but seeks to involve all nations in a coordinated effort to limit global warming to 2°C or less, relative to pre-industrial temperatures, by the end of the 21st century. The agreement also aims to strengthen the ability of all countries to adapt to the effects of climate change. The Paris Agreement requires all Parties to put forward their best efforts through nationally determined contributions (NDCs). This includes requirements that all parties report regularly on their emissions and implementation efforts.

By 2015, 195 countries had signed the Accord and by 2019 184 had ratified the Agreement. The US initially signed the Agreement and submitted an NDC but then in 2018 withdrew from the Agreement. As of the end of 2018, nearly all signatories were falling short of their goals. Of the 32 signatories that account for approximately 80% of the world's greenhouse gas emissions, only 7 had made commitments or efforts that would meet the Paris targets and half were either "critically" or "highly insufficient" in meeting their goals, including the US and China.

Under the Paris Agreement the developed world is to provide financing and technology transfer to the developing world and to assist them with implementation of mitigation and adaptation strategies. Both China and India have indicated they are developing countries and must balance the interests of economic and social development against the interests of climate change mitigation and adaptation. The United States, China and India submitted NDC's in early 2016. A comparison of these primary commitments illustrates many of the key issues facing implementation of effective climate mitigation strategies. (Table 6.7) All of the NDC's used 2005 as the reference point and used 2025 or 2030 as the target date for proposed changes.

In anticipation of the need to provide energy for their rapidly growing economies, China and India defined their carbon emission targets in terms of a reduction in CO_2 emissions per dollar of GDP growth. Such an approach permits these countries to meet their targets with increased carbon emissions as long as the rate of increase in GDP exceeds the rate of increase in carbon emissions. China and India did not set targets for methane and nitrous oxide emissions. In addition, both China and India set targets for non-fossil fuel energy and reforestation, included plans for climate change adaptation, and India estimated implementation costs. The social implications of climate change control policies are featured prominently in both China's and India's NDC's. In contrast, the United States submitted a simple commitment to reduce carbon dioxide-equivalent emissions and outlined the policies and regulations that would be the basis for these reductions. No commitments were made regarding non-fossil fuel energy, reforestation, or energy efficiency. The US did not address climate change adaptation plans or provide any cost estimates. The US included all greenhouse gases and set a target for reducing CO_2-equivalent emissions, which has the desirable effect of lowering radiative forcing. The targets for China and India do not achieve this goal.

Table 6.7: Comparison of NDC targets for China, India, and the United States

Contribution Target	China	India	US
CO_2 emissions/ $GDP 2005 to 2030	Decrease 60-65%	Decrease 33-35%	None stated
CO_2-equivalent emissions 2005 to 2025*	None stated	None stated	Decrease 26-28%
Non-fossil fuel primary energy consumption in 2030	Increase ~20%	Increase ≥40%	None stated
Reforestation by 2030	Increase Forest Stock Volume by 4.5 billion m³	Create carbon sink of 2.5 GtC through forestry	None stated
Energy efficiency	Increase	Not stated	Not stated
Adaptation strategies	Yes	Yes	No
Legal framework/ policies stated	Yes	Yes	Yes
Cost estimate	No	Yes	No

*CO_2-equivalents calculated based on 100-year Global Warming Potential (GWP).

Progress in Meeting NDC Targets

From 2009 to 2019, global CO_2 emissions increased by about 15% or about 1.1% per year.[123] Dramatic increases in CO_2 emissions occurred in China (27%) and India (55%), while the US achieved a 3% reduction in CO_2 emissions over this time period. (Table 6.8)

Table 6.8: Carbon dioxide emissions in gigatonnes from the US, China, and India in 2019 compared to 2009

Country	2009 CO_2 emissions (GtCO₂)	2019 CO_2 emissions (GtCO₂)	Change (%)
US	5.3	5.0	Decrease 3.0
China	7.7	9.8	Increase 27.0
India	1.6	2.5	Increase 55.0
World	29.7	34.2	Increase 15.0

123 British Petroleum Company. (2020). *BP statistical review of world energy*, 69th edition. London: British Petroleum Co.

However, as determined by a change in CO_2 emissions per \$1000 GDP from 2009 to 2019, China achieved a 54.7% decrease, India a 27.3% decrease, Europe a 13.4% decrease, and the US a 36.7% decrease. (Table 6.9) It appears that both China and India had achieved more than half of their proposed 2030 targets at the time they submitted their NDC in 2016. If the Chinese economy increases to \$64.2 trillion in 2030 and China achieves the mid-point in its CO_2 reduction target of 62.5%, then Chinese CO_2 emissions would be 19.4 gigatonnes in 2030, or double the CO_2 emissions in 2019. Emission levels of this magnitude would increase radiative forcing and promote additional global warming.

Table 6.9: Carbon dioxide emissions in kg/\$1000 GDP in the
US, China, and India in 2019 compared to 2009

Country	2009 CO_2 emissions (kg/\$1000 GDP)	2019 CO_2 emissions (kg/\$1000 GDP)	Decrease (%)
US	732	463	36.7
China	3,022	1,370	54.7
India	2,378.5	1,729	27.3
Eurozone	710	615	13.4

The US and Europe achieved about a 10 to 12% reduction in CO_2 emissions per capita from 2009 to 2019 while India and China increased per capita CO_2 emissions by 20% or more. (Table 6.10) The Indian NDC notes that no country has achieved an acceptable level of life expectancy, education, and prosperity (as measured by a Human Development Index greater than 0.9) without per capita energy consumption of at least 4 tonnes of oil equivalent per capita. Assuming both countries have a population of about 1.5 billion in 2030, this level of consumption equates to about 6,000 million tonnes of oil equivalent (mtoe) for both countries. In 2018, China consumed 3,273.5 mtoe and India 809.2 mtoe. Thus, to consume at least 4 mtoe/capita, India would have to install at least 5,200 mtoe and China 2,700 mtoe of low-carbon power generation to avoid dramatically increasing world carbon emissions. This equates to about 0.2 ZJ for India and 0.1 ZJ for China. Assuming the manufacturing resources were available for such an undertaking, the cost over the next 10 years would be \$20-25 trillion for India and \$10-12.5 trillion for China.

Table 6.10: Comparison of per capita carbon dioxide emissions
in the US, China, and India in 2009 and 2019

Country	2009 CO_2 emissions (tonnes/person)	2019 CO_2 emissions (tonnes/person)	Change (%)
US	17.2	15.1	Decrease 12.2
China	5.8	6.85	Increase 18.1
India	1.3	1.8	Increase 38.5
Eurozone	6.2	5.5	Decrease 11.3

Installation of non-fossil fuel primary energy production is another key target for India and China. The US did not include this element in its NDC. Table 6.11 compares the hydroelectric, nuclear, and renewable power consumption in the US, China, and India in 2009 and 2019. During this 10-year period, large increases in renewable power consumption were made in all three countries. The gains in China are especially impressive. While the relative increases were large over this time period, the total power generation in these three countries from non-fossil fuel sources was only 39.96 EJ out of 270.41 EJ of total energy consumption or about 14.6%. In 2019, nuclear, hydroelectric, and renewable energy consumption was 9.0% of total energy consumption in India, 16.7% in the US, and 14.9% in China. It seems highly unlikely India will be able to meet its NDC target of at least 40% non-fossil fuel primary energy by 2030 while China will likely meet its target of a 20% increase in non-fossil fuel primary energy production relative to 2005 production.

Table 6.11: Comparison of non-fossil fuel energy consumption in 2009 and 2019 for the US, China, and India

Country	2009 Power Consumption (EJ)	2019 Power Consumption (EJ)	Change (%)
Hydroelectric			
US	2.56	2.42	Decrease 5.5
China	5.81	11.32	Increase 94.8
India	1.00	1.44	Increase 44.0
Nuclear			
US	7.94	7.6	Decrease 4.3
China	0.66	3.11	Increase 371.2
India	0.16	0.40	Increase 150.0
Renewables			
US	2.39	5.83	Increase 144.0
China	0.52	6.63	Increase 1,750.0
India	0.27	1.21	Increase 348.1
Total Non-Fossil Fuel (Total Consumption %)			
US	12.89 (14.3%)	15.85 (16.7%)	Increase 23.0
China	6.99 (7.2%)	21.06 (14.9%)	Increase 201.3
India	1.43 (6.6%)	3.05 (9.0%)	Increase 113.3

Improving energy efficiency was listed by both India and China as a key strategy in their NDC's. Table 6.12 compares the increase in energy efficiency for the US, China, and India from 2009 to 2019. Energy efficiency is measured by the amount of power needed to produce a dollar of GDP growth. The less power a country consumes to produce a dollar of GDP growth, the more energy efficient it is. In 2019 global energy efficiency was approximately 6.7 MJ/$GDP. The US is much more energy efficient than China and India, reflecting the relative development of their respective economies. All three countries have produced substantial improvements in energy efficiency over the last decade. These improvements will need to continue at 1-3% per year, if targets for global energy efficiency are to be met in 2030.

Table 6.12: Comparison of energy efficiency in 2009 and 2019 for the US, China, and India

Country	2009 Energy Efficiency (MJ/$GDP)	2019 Energy Efficiency (MJ/$GDP)	Increase (%)
US	6.22	4.42	28.9
China	19.11	9.88	48.3
India	16.04	11.87	26.0

Both China and India included targets for increasing carbon removal from the atmosphere through reforestation. Both countries describe their respective plans to increase forest stock volume and create carbon "sinks" to offset carbon emissions. According to the UN Food and Agricultural Organization, the US and India have had stable levels of forestation from 2005 to 2016. The US had 3.05 million km² of forest in 2005 and had 3.10 million km² in 2016. India had 0.68 million km² of forest in 2005 and 0.71 million km² in 2020. China increased its forest cover from 1.93 million km² in 2005 to 2.20 million km² in 2020, an increase of 27 million hectares (66.7 million acres). This level of reforestation is equivalent to 2.733 billion m³ of forest stock volume. China will need to increase its forest cover by about 17 million hectares through 2030 to meet its reforestation target, which seems possible given the effectiveness of its prior programs. It seems unlikely India will meet its carbon sink goal through reforestation and will need to develop other means to do so.

Both China and India discuss methods to adapt to climate change in broad terms. These strategies include improved agricultural productivity, development of agricultural technology to improve crop resilience, improved agricultural practices to lower methane and nitrous oxide emissions, improved water conservation, development of fresh water production technology, city planning to improve energy efficiency, coastal protection from sea water incursion, early warning systems for severe storms and other natural disasters, disaster prevention and mitigation, improved public health, support for low carbon industries, and research and development of new energy or environmental technologies. India estimated its need for at least $2.5 trillion to achieve its goals through 2030, although this is likely an underestimation. The US did not include any mention of adaptation to climate change. Table 6.13 is a partial summary of the various goals and objectives mentioned in these and other NDCs.

Table 6.13: Partial list of strategies, goals, and objectives to meet NDC targets

Strategy	Goal	Specific Objectives
Reduce Greenhouse gas emissions	Increase use of renewable energy	Install Wind, Geothermal, and Solar Energy Units
	Reduce emissions from use of fossil fuels	Reduce Coal/Petroleum use, Increase Natural Gas use
	Increase use of nuclear energy	Improve nuclear reactor technology Increase reactor safety
	Limit methane and nitrous oxide emissions	Improve agricultural practices Maintain pipelines and wells to prevent leaks
Increase photosynthetic carbon fixation	Improve land use	Decrease deforestation Increase reforestation
Increase energy efficiency	Improve the electrical grid	Reduce energy loss Improve efficiency
	Develop and deploy methods for local energy production & storage	Improve battery technology Improve hydrogen production technology Reduce the cost and improve the safety of fuel cells
	Increase vehicle fuel economy standards	Build Hybrid vehicles Promote use of electrical vehicles and mass transportation
	Increase energy efficiency in the construction sector	LED lighting Engineering improvements
Capture and store or utilize carbon emissions	Capture and store or utilize atmospheric or industrial carbon emissions	Deep well or deep ocean injection

The Kyoto and Paris agreements were negotiated to be acceptable to the world's political leadership at the time of adoption. Since 2000, the amount of installed wind and solar nameplate capacity has increased dramatically and the amount of "clean" energy produced has increased 10 fold. The amount of investment in renewable energy has increased worldwide to about $350 billion annually but hasn't significantly increased the percent of global energy production from renewable resources. Solar, wind, and biofuels accounted for about 220 terawatt hours (TWh) of 112,500 TWh of total global energy produced in 2000 (0.2%) and about 2000 TWh of the 157,135 TWh of total global energy produced in 2018 (1.2%). If solar and wind energy are deployed at about 150 GW nameplate capacity per year for the next 30 years and we assume the capacity factor increases from

0.25 to 0.3, then in 2050 total production from solar, wind and geothermal energy will reach about 14,200 TWh, a sevenfold increase from 2018 requiring an investment of over $11 Trillion over 30+ years. If total energy production in 2050 is 0.8 ZJ or 222,222.2 TWh, solar, wind, and geothermal sources will represent only about 6% of world energy production. While the increase is a major gain, it falls far short of the amount necessary to appreciably offset fossil fuel energy production.

The Path Forward

Population and GDP growth drive many of the anthropogenic causes of greenhouse gas emissions. Power stations, production of fossil fuels, and agriculture account for more than 50% of greenhouse gas emissions. Industrial processes, transportation, residential, commercial, and waste disposal account for the remainder. Efforts to mitigate future greenhouse gas emissions must be developed based on projected future population growth, energy consumption, changes in the sources of the energy, and the availability of new technology. Planners must make a set of assumptions about future conditions in order to prioritize and estimate the impact of interventions designed to stabilize greenhouse gas emissions.

The following example illustrates the magnitude of the reduction in greenhouse gas emissions that could be achieved using existing methods combined with new technology. It is not a proposed course of action; it is designed to illustrate the dynamics of change. The values in the tables are approximations and would require formal analysis to determine accurate estimates. The assumptions are shown in Table 6.14. The example assumes population control programs are implemented in parallel with carbon mitigation programs.

World population is assumed to increase to 9 billion people in 2050 instead of 9.6 billion. Global population then declines through the end of the century to 8 billion compared with 10-12 billion without additional population control programs. Energy efficiency is assumed to increase by 35-40% by the end of the century, which is an ambitious goal and will require new technology and effective energy conservation programs. The total energy requirement through 2050 is 20 ZJ and the total energy requirement through 2100 is 50 ZJ. These assumptions are aspirational; by no means would they be easy to accomplish or likely to be achieved. They are a "good case" or an "upside" scenario.

Table 6.14: Assumptions

Assumption	2019 to 2050	2051 to 2100
World Population	9 Billion by 2050	8 Billion by 2100
Annual GDP Growth	2-2.5%	1-1.5%
Energy efficiency	Increase 35-40%	No further Increase
Annual World Energy Consumption	0.75-0.8 ZJ in 2050	0.6-0.65 ZJ in 2100
Total Energy Requirement	20 ZJ	50 ZJ

Table 6.15 illustrates one possible approach to producing 20 ZJ of energy through 2050 while lowering carbon emissions. In this example, oil use decreases by 1% per year from 35 billion barrels in 2018 to 23 billion barrels in 2050, natural gas utilization goes up by 1% per year from 135 billion mcf to 175 billion mcf, and coal use goes down by 3% per year from 7 billion tonnes to 2 billion tonnes. This mix of fossil fuels would generate about 380-400 EJ of energy per year in 2050. Nuclear, hydroelectric and biomass would generate about 75 EJ and wind, solar, and geothermal power plants would generate about 310 EJ (about 40%) of global energy production. In 2050, energy production would be 750-800 EJ. This level of energy production would be at the low end of current 2050 energy consumption projections.

Table 6.15: Sample path to meet energy demand in 2050 and lower carbon emissions

Fuel	2017 Production (EJ)	Change/year to 2050 (%)	Total energy produced to 2050 (ZJ)	Energy produced in 2050 (EJ)
Oil	194	Decrease 1.0%	5.5	140
Natural Gas	132	Increase 1.0%	5.1	182
Coal	156	Decrease 3.0%	3.3	59
Nuclear	30	Increase 1.0%	1.0	40
Hydroelectric	40	Unchanged	1.2	40
Biomass	4	Unchanged	0.1	4
Wind/solar/ geothermal	35	Increase to 320 GW/year*	5.0	310
Total	591		21.2	775

*Nameplate capacity, 30% capacity factor.

This approach reduces carbon emissions from 9.9±0.5 gigatonnes of carbon (GtC) in 2018 to an average of approximately 7.8 GtC per year through 2050. (Table 6.16) Emissions from fossil fuels in 2050 are ~6.7 GtC. This example doesn't rapidly or profoundly alter energy markets in the near term, but it significantly reduces the amount of carbon emissions through 2050.

Table 6.16: Calculation of carbon emissions from sample path

Fuel	Giga-Joules/Unit	Amount used through 2050 (Trillion)	Total carbon produced (Gigatonnes)	Average Carbon emissions/year (Gigatonnes)	Carbon Emissions in 2050 (Gigatonnes)
Oil	6/barrel	0.94 barrels	105	3.2	2.7
Natural Gas	1.1/mcf	4.6 mcf	74	2.3	2.4
Coal	30/tonne	0.11 tonnes	76	2.3	1.5
Total				7.8	6.7

Annual methane and nitrous oxide emissions contribute to an increase in atmospheric CO_2-equivalent concentrations. Only about 50-60% of these emissions are anthropogenic. Cattle account for about 40% of methane emissions. The remaining 10-20% comes from fracking, leaks from natural gas wells or pipelines, and other sources. Soil is the main source of natural methane emissions. Similarly, farming is the main source of anthropogenic nitrous oxide emissions. China, the US, and India are the largest source of methane emissions. Reducing anthropogenic methane and nitrous oxide emissions must be accomplished through improved agricultural and industrial practices and development of new technology.

In order to stabilize atmospheric greenhouse gas emissions by 2050, deforestation will need to be stopped completely within the next 10 years, effectively adding photosynthetic carbon fixing capacity of 1.3±0.5 GtC per year. Ocean uptake of carbon is assumed to continue at 3.0±0.5 GtC per year. A combination of reforestation and growing algae could be used to increase photosynthetic carbon fixation. Trees can fix about 2-3 tons of carbon per acre per year. Planting 100-200 million acres (40 to 80 million hectares) of rapidly growing trees worldwide in the next ten years could remove about 0.3 to 0.6 GtC per year. This planting would expand the world's forests by 2% and represent a reforestation rate that is about double the rate at which China reforested in the past decade. Microalgae have the highest photosynthetic capacity on Earth. Algae can fix about 50 to 75 tons of carbon per acre if they are directly fed carbon dioxide under proper conditions. Building a few thousand 2,000-acre algae farms globally could utilize about 0.2 to 0.3 GtC derived from flue gas per year. Finally, directly removing carbon dioxide from the atmosphere using gas adsorption technology and then either storing or utilizing the carbon dioxide could remove about 0.5 GtC per year. Ideally, carbon dioxide adsorption could be used to provide carbon dioxide to the algae farms. Table 6.17 summarizes the resulting estimated 2050 carbon balance.

Table 6.17. Change in estimated carbon balance in 2018 and 2050

Method	2018 (GtC)	2050 (GtC)	Change in Emissions (GtC)
Burning fossil fuels	+9.9	+6.7	−3.2
Deforestation	+1.3	0.0	−1.3
Net Ocean carbon uptake	−3.0	−3.0	Unchanged
Current net land uptake	−2.0	−2.0	Unchanged
Reforestation	0.0	−0.5	−0.5
Algae farms	0.0	−0.3	−0.3
CO_2 Adsorption	0.0	−0.5	−0.5
Total Net Emissions	+6.2	+0.4	−5.8

These steps would reduce fossil fuel atmospheric carbon emissions from 9.9 GtC per year in 2018 to 6.7 GtC per year in 2050. The Earth would not be carbon neutral in 2050 but could achieve equilibrium after 2050 through continued decreases in energy consumption, improved energy efficiency, carbon removal strategies, and additional deployment of carbon neutral energy sources. Even with these aspirational improvements, global surface temperatures would still likely increase more than 2.0°C by 2100.

This example is simplistic and idealized. It demonstrates that carbon emissions can be significantly reduced if population growth is controlled, energy efficiency improves, and energy consumption is limited to 750 to 800 EJ in 2050. This level of energy consumption is approximately 2 tonnes of oil equivalent (toe) per person, which is 8% higher than in 2018 but falls far short of the goal of achieving a global energy intensity of 4 toe/person. If India and China aspire to this level of energy intensity by 2050, just those two countries would consume 500 EJ per year. In 2018 the OECD, which does not include China and India, consumed about 240 EJ of energy. Even if the OECD energy consumption remained constant for the next 30 years, China and India could not achieve their energy intensity goals while limiting world energy consumption to 800 EJ or less.

Goal 3: Increase investment in new technology and/or improve existing technology.

As the pace of life has accelerated, we have become accustomed to rapid change. These changes have been largely related to the development of new technology in agriculture, medicine, communication, and information management. In fact, our faith in the ability of mankind to solve problems with technology is rather remarkable. However, it has been a very long time since mankind has encountered profound and sustained global changes in the environment, widespread scarcity of essential resources, or mass human migration.

Faced with these challenges, scientists, engineers, and entrepreneurs have pursued new technology and new business opportunities. Table 6.18 lists a few of these new technologies. There is no doubt new technology should be an element of any reasonable plan to mitigate climate change. However, many of these new technologies are still in development, are likely to be costly, may not be scalable to meaningful levels of operation, and may use more energy than they save. Failing to reduce population growth and global energy demand and relying on new technology to mitigate the carbon emissions from continued GDP growth is a very risky approach that has a high likelihood of failure.

A less risky approach is to continue to improve existing low carbon energy technology, such as wind and solar technology, and to further invest in improving geothermal and tidal technology where tectonic activity or local tides permit efficient operation. Of course it is always possible to fantasize about the development of new, transformative energy technology that safely produces vast amounts of cheap, clean energy at an affordable cost and that can be deployed at very large scale over a few decades, including in the developing world.

Table 6.18: New technologies for reducing carbon emissions

New Technology	Comment
Carbon Dioxide Capture	Capture of CO_2 from flue gas or the atmosphere using adsorbents
Carbon Dioxide Storage	Purification and compression CO_2, deep Earth or ocean injection of compressed carbon dioxide; alternatively injection of algal biomass produced from captured CO_2
Carbon Dioxide Utilization	Conversion of CO_2 or biomass into commercial hydrocarbons
Battery technology	Increased energy density, lower cost, longer life; solid state batteries
Fuel Cells/Hydrogen	Reduced cost, increased safety, improved hydrogen production and storage
Clean Coal	Various coal gasification and flue gas treatment technologies
Ocean seeding	Iron salts or other chemicals used to promote phytoplankton growth and photosynthesis in the oceans
Microgrids	Localized grouping of solar and wind farms; combined with high capacity factor, local generation and batteries to manage the microgrid electrical load
Wave and Tidal	Tidal or wave kinetic energy used to power turbines
Algae	Microalgae farms to increase photosynthesis
Energy storage (other than batteries)	Surplus energy production stored as heat or other potential energy sources
Increase Earth's albedo	Orbiting mirrors, injection of sulfate aerosols into the stratosphere, cloud seeding, painting roofs and other large surfaces white

Goal 4: Develop methods to finance and implement programs by nation or region.

Economic incentives and disincentives can be used to promote lowering carbon emissions. One approach is to tax carbon emissions and use the tax revenues to promote low carbon energy production. Another approach is to limit carbon emissions through "cap-and-trade" systems. In these systems government agencies set annual carbon emission "allowances" for industries that have historically produced high levels of carbon dioxide or other greenhouse gas emissions. These industries include power generation, oil refineries, steel, iron, aluminum, cement, paper, glass, and aviation.

By setting these allowances a government is setting a cap on greenhouse gas emissions from these industries, which often account for the majority of greenhouse gas emissions in a country. A government can then allocate these allowances to individual companies in a variety of ways that may or may not involve payment for an allowance. If the greenhouse gas emissions from a company exceed their allowance, the company pays a fine to the government. If a company does not use all of its allowances, it can sell them on secondary markets maintained by the government. The secondary markets may be regulated in various ways to be certain the supply of allowances is consistent with demand and that prices don't become distorted. Under this system companies have an incentive to increase their energy efficiency so they have excess allowances they can sell in secondary markets. The European Union established such a system in 2005. China, South Korea, several US states, and other countries or regions have also implemented cap-and-trade programs.

In practice, these cap-and-trade systems have some problems. Governments are often generous with free allowances at the outset to aid in gaining acceptance of new regulation. This practice results in excess supply of allowances in the secondary market and low prices for allowances. Governments have responded by instituting a reserve system and removing excess allowances from the secondary market. They also conduct periodic auctions that have resulted in increasing prices. In Europe, the cap-and-trade system has not reduced the use of coal by the electrical power industry, but the program is credited with reducing, or at least stabilizing, carbon emissions.[124] While economists and politicians continue to debate the value of cap and trade, it does appear that coercive economic programs such as a carbon tax are likely to become increasingly prevalent as a means to reduce carbon emissions.

It will be very difficult to control carbon emissions when global population and GDP are increasing. It will require major changes in the energy sector, although several of these changes have already begun. Countries dependent on oil for revenues such as Saudi Arabia,

---

124   Bayer P., Aklin, M. (2020) The European Union Emissions Trading System reduced CO_2 emissions despite low prices. Proc Natl Acad Sci: 117(16): 8804–8812.

UAE, Iran, and Russia would likely resist limitations on oil production but would benefit from increased natural gas production. India and China would need a major overhaul of their electrical power generation industry, which is currently heavily dependent on coal. Major coal exporting countries, such as Australia, would likely resist such changes. Halting deforestation, especially in countries such as Brazil and Central Africa has been an elusive goal. Financing deployment of wind and solar farms on a massive scale, major new reforestation programs, and the deployment of new algae and gas adsorption technology would be risky. There will be strong political resistance to change.

Short-term economic and political interests often prevail over the interest of mitigating long-term risks. Individual countries and especially individual developing countries want to increase per capita GDP. To achieve this goal they need to increase their energy intensity. Unfortunately, the short-term benefits of this strategy will be outweighed by the long-term consequences. A great deal of vision and political courage will be required to make the changes necessary to mitigate climate change, and these qualities are in short supply. The Paris Agreement does not achieve this goal.

This example illustrates several of the challenges that must be overcome to achieve the mission of creating a sustainable world in the second half of the 21st century. Multiple strategies will be required, including population control and substantial increases in energy efficiency. An immediate doubling of the rate of wind and solar power deployment does not, by itself, materially alter the prospects for global warming. Fortunately, there are many available strategies and technologies. Whatever approach is adopted it must be carefully designed, managed, and monitored. It is unlikely the needed steps will be achieved by a voluntary program that is based on individual countries meeting self-imposed targets. With such an approach, short-term economic and political interests will prevail over the long-term interest of mitigating climate change. Without the leadership of China, the US, and India, meaningful progress is unlikely.

Natural Resource Depletion

Natural resources are either renewable or non-renewable. Resources are renewable if a natural process can replenish losses in a timeframe that is meaningful for humans. Renewable resources include wind and solar energy, fresh water, and forests. Non-renewable resources include fossil fuels, minerals, and certain groundwater aquifers. Arable land is an essential natural resource that can be depleted, although arable land can be replaced through deforestation or conversion of agricultural land. Depletion of non-renewable natural resources is dependent on the size of the reserves and the rate of consumption.

Consumption is determined by several factors, including population size and GDP growth, efficiency of use, the development of new technology, and the cost of production. The rate and extent of depletion of non-renewable resources and the supply of renewable resources varies from one geographic region to another. If supply is low relative to consumption, the availability of a natural resource may be stressed, or the resource may become scarce or depleted. Non-renewable resources with low supply relative to consumption must be conserved to avoid scarcity or depletion. Renewable resources may also become scarce if consumption exceeds supply over a time frame that exceeds the rate at which the resource can be replenished.

Renewable resources require conservation as well as non-renewable resources. Conservation methods include recycling, reducing waste or unnecessary loss, preventing contamination or pollution, and developing alternative resources, if possible. Government regulation and changing human behavior are means to alter the rate of consumption of natural resources and especially the rate of consumption of non-renewable resources.

If current projections materialize, world population in 2050 will be 9-10 billion people and will be 10-12 billion people in 2100. In 2050 global GDP will be ~$182 trillion (measured in 2010 $US) and in 2100 >$450 trillion. Assuming a 35-40% increase in energy efficiency, global energy consumption in 2050 is likely to be between 0.75 and 0.9 ZJ. By 2100 world energy consumption will likely exceed 1.1 ZJ per year. If so, the world will need to produce 22 to 26 ZJ of energy between now and 2050 and 70 to 80 ZJ of energy by the end of the century. Current recoverable reserves of oil and natural gas total about 11 ZJ for oil and 13 ZJ for natural gas. Recoverable reserves include proven reserves and reserves that are known but not currently under production. These reserves have a high likelihood (>90%) of full recovery. If we include unconventional sources and additional undiscovered reserves that may exist but will require new technology and/or much higher market prices to make recovery economical, the estimated total reserves increase to approximately 18ZJ for oil and 16 ZJ for natural gas. These additional reserves, 7 ZJ (or 1.15 trillion barrels) for oil and 3 ZJ for natural gas (2.8 trillion mcf), have a low likelihood of full recovery (less than 10%).

Petroleum and natural gas are used for much more than generating energy. They are important raw materials for producing petrochemicals such as methanol, formaldehyde, ethylene, propylene, butadiene, benzene, toluene and xylene. For example, natural gas can be converted to methanol or to synthesis gas (syngas), which is a mixture of hydrogen and carbon monoxide. Syngas is a versatile raw material that can be used for many purposes, such as making ammonia for fertilizer. In the US, more than 15% of oil and about 10% of natural gas production is used to produce petrochemicals. Only about 56% of oil is refined into gasoline and only about 40% of natural gas is used for generating electricity. Examples of end products produced from petrochemicals include pharmaceuticals, cleaning products, rubber, cosmetics, insulation, lubricants, plastics, fabrics, animal feed, and paint. Demand for petrochemicals is expected to increase substantially during the remainder of this century.

Under almost any plausible energy scenario, oil and natural gas demand is likely to increase this century, especially if coal use is reduced and deployment of renewable energy fails to keep pace. Non-energy demand for oil and natural gas is also likely to increase, especially if demand for methanol, ammonia, and other chemical raw materials increases. Petrochemicals are set to account for more than a third of the growth in world oil demand by 2030, and nearly half of the growth by 2050.

Unless new technologies are developed to substitute for these raw materials, preservation of non-renewable oil and natural gas resources for future generations will be needed. The demand for petrochemicals and energy through 2050 will likely support further growth of the oil and gas industry, even if use for energy production is reduced in order to lower carbon emissions. However, if demand for petrochemicals increases beyond 2050, use of oil and natural gas for energy production will need to decrease substantially in order to conserve oil and natural gas reserves for the 22nd century. Without such efforts, oil and natural gas reserves will likely become depleted in the 21st century; even if oil and natural gas are much more abundant and unconventional reserves are economically recoverable with new technology.

British Petroleum in its 2019 Statistical Review of World Energy estimated the 2018 proven reserve to production ratio for both oil and natural gas to be 50 years each, assuming constant 2018 production rates. Even if total oil and natural gas reserves are twice as large as 2018 proven reserves, increased production will deplete total reserves by the end of the century.

There are very large reserves of coal on Earth relative to the expected rate of future consumption. Proven reserves are 1.05 trillion tonnes of coal or ~31 ZJ. Undiscovered reserves could be as large as 0.7 trillion tonnes. Consumption was about 7.9 billion tonnes in 2018 and has been relatively stable (±2%) for the past few years. About 67% of global coal production is used for generating electricity and the remainder goes for industrial purposes, especially production of steel.

The proportion of electricity produced from coal varies between countries. In China coal accounts for 68% of electricity generation, in India 75%, and in the US just 31%. In the past few years reductions in coal use in the US and Europe have been offset by increased consumption in India, Indonesia, and Turkey. In 2017 US coal consumption decreased by almost 20,000 tonnes while India increased consumption by 40,000 tonnes and China by 10,000 tonnes. China has increased coal consumption but has decreased the proportion of electricity generated from coal because the country has been investing in nuclear and renewable sources.

Many countries with large coal reserves increase use of coal because it is cheaper than alternative energy sources or because they want to avoid or lessen their dependence on foreign oil and natural gas for energy production. The global 2018 reserve to production ratio (R/P) for coal is 132 years; for China, the R/P ratio is 38 years; for India, 132 years; and for the US, 365 years. China currently imports large amounts of coal from Australia, Indonesia, Mongolia, and Russia and can be expected to continue to do so. If global carbon emissions are to be stabilized and eventually reduced, use of coal for generating electricity must be reduced dramatically, unless effective and affordable flue gas carbon capture technology is developed and deployed on a very large scale.

Carbon capture technology has been developed and used in the industrial gas business for years. Generally, the technology involves passing a gas containing carbon dioxide over a catalyst that adsorbs the carbon dioxide. After the catalyst is saturated with carbon dioxide, it is heated or otherwise treated so as to cause the carbon dioxide to be released from the adsorbent. The CO_2 is then collected and concentrated or liquefied. A high concentration of carbon dioxide in the afferent gas reduces the cost and increases the efficiency of the process. Carbon capture technology has been deployed at the million-ton scale, but large amounts of energy are required, and it is currently uneconomical for flue gas or atmospheric carbon capture. There are many research and development projects underway worldwide to improve the performance and lower the cost. These projects include a search for better catalysts and development of electrochemical or thermochemical methods to increase process efficiency.

Once carbon dioxide has been captured, it can be concentrated and compressed into a liquid for transportation or storage. Injection of large amounts of CO_2 into aquifers or subterranean rock formations has been evaluated in several demonstration projects. Some laboratories have even studied the feasibility of mineralizing CO_2. Carbon dioxide can also be converted into commercial products using various chemical or biological systems. Chemical conversion to methanol and biological conversion to biomass through photosynthesis are examples of CO_2 utilization methods. Like carbon capture, improved performance and lower cost are essential before these methods can be deployed at large scale. Many of these methods likely depend on a carbon tax or cap-and-trade system to make them sufficiently profitable in order to attract investment.

The coal industry is investing in methods to reduce carbon emissions from electricity generation and in methods to use coal for other commercial purposes. These methods depend on converting coal into syngas and then using the syngas to power a combined cycle power plant or using the syngas as feedstock for producing petrochemicals. Syngas is a mixture of carbon monoxide and hydrogen. Coal gasification begins with combining crushed coal with steam at high temperature and pressure in a low oxygen atmosphere. The result is production of carbon monoxide and hydrogen gas. There is always some combustion, so some carbon dioxide is produced as well. Impurities in the coal also produce hydrogen sulfide (H_2S), ammonia, cyanide gas, and mercury in the gas phase and some heavier minerals can be removed as solids. The CO_2 and H_2S can be removed by passing the gas over an adsorbent and the ammonia, cyanide, and mercury can be removed by hydrolysis or by cooling the gas. The resulting mixture of carbon monoxide and hydrogen is relatively pure. The syngas can be further combined with water to generate carbon dioxide and even more hydrogen. The hydrogen can then be burned to produce heat for turbines that generate electricity.

Coal gasification can be used in Integrated Gasification Combined Cycle (IGCC) power plants to generate electricity. In an IGCC power plant the hydrogen from syngas drives a gas turbine. The heat generated is captured and used to drive steam turbines for additional electricity generation. Carbon dioxide can be removed, along with other impurities, prior to combustion of the hydrogen. Heat exchangers are used to capture process heat and increase the efficiency of the process.

Several IGCC power plants have been built and operated at demonstration scale. High construction and operating costs are major barriers to further deployment. Further process improvements are necessary to demonstrate effective carbon capture and removal of toxic impurities. This technology is controversial. Critics point to the poor performance, toxic impurities produced, low efficiency of carbon removal and high cost. They advocate for using scarce capital to invest in renewable energy and not in "clean" coal technology. Proponents recognize the need for further process improvements and cost reduction but point to the value of being able to continue to use the vast global reserves of coal for power production without damaging the environment.

Global reserves of coal will not be depleted this century. However, coal reserves will likely be depleted in some geographic regions, such as China and possibly India, if coal use continues to increase. The US will certainly continue to have sizable reserves long into the future. The future of the coal industry is dependent on improving the efficiency and reducing the cost of IGCC and other "clean coal" technologies or the development of low cost flue gas carbon capture and storage or utilization technology.

Power generation from biomass and hydroelectric sources are unlikely to increase and may decrease because of climate change. In 2018 nuclear power plants only supplied about 11% of global electricity demand and about 1.5% of global total energy demand. There are 438 operating nuclear power plants and 62 nuclear plants under construction. Unfortunately, the world's nuclear reactors are also aging, and many have reached or exceeded their design lifetimes. The growth of the industry in the next 50 years will be driven primarily by the rate at which existing plants are decommissioned and the amount of investment available for new plant construction. The amount of time and money that must be invested to build a new nuclear power plant is high and there are alternative less expensive low carbon energy sources, such as wind, solar, and geothermal, that compete for the financial resources available for new energy production.

Existing uranium ore reserves recoverable at less than $250,000 per tonne are about 8.4 million tonnes. Global annual consumption is about 86,000 tonnes, resulting in a reserve to production ratio in 2018 of about 100 years. Economical U_{235} will likely be scarce by the end of the century. However, uranium is plentiful on Earth in very low concentrations and large recoverable reserves of good quality ore may be added due to melting of ice caps and thawing of glaciers.

Arable Land

Only about 6-10% of the Earth's total land surface area is arable, or about 3.5 billion acres (1.4 billion hectares). Per capita arable land needs are 0.25 to 0.75 acres. Assuming continued improvement in agricultural productivity and reduction in food waste, the amount of arable land on Earth should be able to support as many as 10-14 billion people. However, by the second half of the 21st century the Earth will most likely have reached its full capacity to support human habitation. Global warming, droughts, desertification, urbanization, and pollution could reduce this number below 10 billion unless steps are taken to conserve arable land and use it optimally. In the last 20 years about 370 million acres of arable land have been added through deforestation and irrigation and about 360 million acres of arable land have been lost due to desertification, urbanization, pollution or sea level rise. As a result, the total amount of arable land on Earth has not changed significantly in the last 20 years. The amount of agricultural and arable land is not evenly distributed across the major geographic regions of the world. Asia and Africa have the largest amount of land but have much less arable land per capita than Europe, Australia, North America, and South America.

The amount of arable land needed in a country or region is determined by agricultural productivity, the size of the population, the amount of food wasted, and by dietary practices, such as the amount of protein and fat consumed. Improvements in agricultural productivity reduce the amount of arable land needed. There have been dramatic improvements in agricultural productivity over the last 50 years, driven by improved farming practices and technological advances in crop science and farm machinery. Most of these improvements have occurred in the developed world, reducing the demand for arable land. Unfortunately, many of these improvements have not been translated effectively into the developing world.

Most of the world has enough arable land to support projected population growth through 2050, especially if people in the developed world eat less meat and dairy products and agricultural productivity in the developing world continues to improve. Africa does not have adequate arable land to meet future needs, even if the amount of arable land is increased by 50% through deforestation or irrigation and fertilization of agricultural land. Because of the expected increase in population in the next 30 years, the current amount of arable land in Africa is only about 50% of the arable land needed in 2050.

African nations need to redouble efforts to ensure best agricultural practices are being used if they are to have any prospect of meeting the dietary needs of their population. Investment, education, training, seed, and equipment will be needed in large quantities. To meet the shortfall, African countries will need to increase agricultural investment and crop productivity, decrease food waste, irrigate and fertilize pastureland to convert it into arable land, convert forested land to arable land, and import grain and cereals in large quantities from regions of the world with excess production capacity. Africa cannot afford to export agricultural products with a large water footprint. If climate change in Africa leads to desertification, droughts, or other environmental conditions not suitable for agriculture, the prospects for producing adequate amounts of food for the continent look bleak. If climate change should reduce crop productivity, especially in Nigeria, Sudan, Ethiopia, and Tanzania, it is unlikely that sufficient arable land will be available to support the projected population growth in Africa.

Fertile soils are limited and essentially non-renewable, at least on human time scales. It is crucial to protect available soil resources from degradation. There are a number of steps that can be taken to improve agricultural productivity and to conserve arable land. Developing more heat, draught, and disease resistant crops, optimizing use of fertilizers and irrigation, and preventing or ameliorating pollution or depletion of soil nutrients can improve agricultural productivity. Additional steps include land reform to promote farming as a way of life, improved infrastructure to make farming commerce more efficient and reduce food waste, and use of information technology to provide farmers with good data on which to base their decisions or to guide farming practices, such as when to plant and when to harvest. The impending food crisis in Africa is a global problem. The social, political and economic consequences of widespread famine are substantial. Conflict, mass migration, pandemics, and increased infant mortality could ensue.

Fresh Water

Fresh water is a finite but renewable resource. There are about 39.8 million km^3 of fresh water on Earth but 29.2 million km^3 of it is in ice caps, glaciers, and permanent snowfields. Most of the remainder is in subterranean aquifers. Rivers and lakes contain about 93,000 km^3 and the atmosphere contains about 13,000 km^3. The distribution of freshwater on

Earth is uneven. Europe, Africa, and South America have the largest reserves of fresh groundwater. Annual rainfall over land is about 119,000 km³. Most of the rainfall occurs in South America, Central Africa, and South Asia. Most rainfall evaporates or runs off into the oceans. Only about 4,500-5,000 km³ of rainfall per year is used for agriculture. Humans remove about this same amount each year from groundwater for agriculture, industrial, or domestic needs. Domestic needs account for less than 15% of freshwater withdrawals. In the developed world most freshwater is used to generate electricity while in the developing world it is used for irrigation.

The amount of accessible, renewable freshwater available per capita can be used to determine the adequacy of freshwater resources. Water stress starts when the water available in a country or region drops below 1,700 m³/year per person. When the water available in a country or region drops below 1,000 m³/year per person, water scarcity is experienced. Absolute water scarcity is considered for countries with less 500 m³/year per person. Some 50 countries, with roughly a third of the world's population, experience water stress, and 17 of these extract more water annually than is recharged through their natural water cycles. Groundwater provides about 90% of the freshwater withdrawals used each day in the world so the health of the groundwater resources is vital to meeting the needs of civilization. Groundwater resources in the southwest of the US, in the Middle East, and in central and south Asia are currently under stress.

The water footprint is the total volume of freshwater used to produce the goods and services consumed by the nation, business, or industry. The water footprint includes the amount of freshwater used in production and the amount of freshwater contaminated in the process. The global per capita water footprint was 1,385 m³/year (Range: 550-3,800 m³/year). This amount includes water for agriculture (92%), industrial purposes (5%), and domestic use (3%), including wastewater.

By 2050 population growth and climate change will either exacerbate or induce water stress and scarcity in several regions of the world. India and South Asia will experience water scarcity and the Middle East and North Africa will experience absolute water scarcity. Sub-Saharan Africa and Northwest China will experience water stress. The rest of the world would appear to have adequate water resources, unless climate change produces more extreme conditions than currently expected. As predicted several years ago, clouds have moved to higher latitudes and become denser over lower latitudes. Regions in the mid-latitudes have experienced a reduction in precipitation and cloud cover. Since 2002 NASA has measured the Earth's land-based water resources in its AQUA Earth-observing satellite mission. The Middle East, South Asia, North Africa, and the southwest of North America are losing groundwater and surface water at an alarming rate. These results suggest deserts in Africa, Asia, and North America are likely to get bigger and many regions of the world with marginal current supplies of freshwater are likely to experience scarcity by

2050. Should sub-Saharan Africa experience additional water scarcity it is unlikely this region will be able to support population growth.

For water-scarce countries such as in North Africa and the Middle East, it is crucial to recognize their dependency on external water resources and to develop foreign and trade policies that ensure a sustainable and secure import of water-intensive commodities that cannot be produced domestically. There are a number of steps that can be taken to conserve water resources. In the developing world most freshwater is used for agriculture, so improving irrigation methods is a critical step in reducing freshwater demand. Water catchment and storage is another strategy that can be effective in regions where annual rainfall occurs sporadically during the year. Desalination is an alternative in countries with the energy and financial resources to support large scale operations, although solar powered desalination may be a new cost-effective strategy. Recycling is an effective strategy for producing non-potable freshwater from wastewater. Preventing pollution and inefficient or unnecessary use of freshwater is also important. Sewage treatment and removal is especially important in highly populated regions. Ensuring piping and water infrastructure can deliver freshwater without contamination or leakage is essential. Finally, information technology can be used to guide water usage, improve water use efficiency, and prevent unnecessary losses.

Summary

It seems clear that mankind must do more than it has during the last 30 years if we are to build a sustainable world in the second half of the 21st century. In order to achieve this mission by 2050, the following goals must be met:

- First, we must develop plans to accelerate the current decline in the population growth rate in those countries where the growth rate exceeds the replacement rate. The long-term goal should be to eliminate positive population momentum and maintain the world population at 9±0.2 billion in 2050.
- Second, we must eliminate the global net increase in atmospheric carbon emissions as soon as possible.
- Third, we must increase investment in new technology and/or improve existing technology that will:

  ○ improve energy efficiency,
  ○ reduce carbon emissions or atmospheric carbon concentrations,
  ○ develop alternatives for producing petrochemicals,
  ○ produce fresh water or protecting ground water reserves,
  ○ improve crop resistance or increasing crop productivity,
  ○ promote adaptation to global warming.

- Finally, we must develop practical and effective methods to finance and implement programs by nation or region.

Lessening population growth is essential to mitigating climate change and natural resource depletion. There are a number of direct population control strategies: 1) strategies that interfere with reproductive biology, 2) strategies that change human behavior, and 3) strategies that directly or indirectly coerce lower total fertility rates. Behavioral and social strategies can be effective in reducing total fertility and population growth rates. These strategies include: 1) improving maternal and child health care, 2) providing family planning services, 3) educating men and women through at least secondary school and including sex education in the curriculum for all students, and 3) promoting gender equality, including actively assisting women who want to enter the workforce. Combining reproductive strategies with behavioral strategies can be highly effective. Promoting abstinence is an ineffective strategy, especially when used alone. Coercive strategies include taxes, subsidies, rebates, or credits and denial or provision of essential services, depending on the number of children in a household. Coercive strategies promote one or two child families and discourage having three or more children.

The major national or international efforts to control population growth are: 1) the UN Population Action Plan, 2) individual national governmental programs, and 3) non-governmental organizations. Urbanization and economic development are associated with decreased total fertility and population growth rates. Population control programs work when managed by stable governments, when non-governmental organizations are actively engaged, and when coercive methods are judiciously employed.

Population momentum has a major impact on population growth and there is a high level of positive population momentum in the developing world. In order to mitigate climate change and natural resource depletion, world population should not exceed 9 ± 0.3 billion in 2050-2060 and should be reduced to less than 8.5 ± 0.5 billion by 2100. There are 10 African and 4 Middle Eastern countries that will account for more than 90% of projected global population growth through 2100. The total fertility rate in these 14 countries must be lowered to replacement levels or below.

Sub-Saharan Africa is a patriarchal society that traditionally places a high social premium on women having many children. Vigorous state-led population control programs in Africa that promote monogamy, provide family planning programs, distribute birth control, and educate both men and women through secondary school have been effective.

Involvement of religious leaders in promoting family planning and use of contraceptives is essential, especially in Islamic countries. There are four Moslem countries with high total fertility and population growth rates: Egypt, Iraq, Pakistan, and Algeria. Low population growth in Islamic countries requires strong support from religious leaders, stable government, and use of behavioral as well as reproductive strategies, especially providing secondary education and economic opportunity for women.

India must reduce the total fertility rate in the Moslem community without increasing religious antipathy between Moslems and Hindus. These population control programs should be equitable and justly coercive. There are several countries that are experiencing low population growth or declining populations. Population decline can have advantageous or disadvantageous consequences. The developed world can use immigration policies to attract workers from the developing world to maintain or grow its labor force.

By 2050 the global mean surface temperature will likely increase 1.6°C to 2.2°C relative to surface temperatures in 2019, assuming atmospheric greenhouse gas concentrations reach 650 ppm of CO_2-equivalents. By the end of the century the rise in the global mean surface temperature will likely be an additional 1.8-2.2°C if atmospheric concentrations of greenhouse gases reach 1100 ppm of CO_2-equivalents. It is doubtful that such an environment would be able to sustain a human population of 10-12 billion. The cause of global warming is a rise in atmospheric concentrations of greenhouse gases (carbon dioxide, methane, nitrous oxide, and chloroflurocarbons), which are caused primarily by human activity. Greenhouse gas emissions can be stabilized and then reduced by: 1) decreasing the use of coal and oil; 2) increasing use of natural gas; 3) increasing installation of wind, geothermal, and solar power; 4) increasing use of nuclear power; 5) improving agricultural and industrial practices to reduce methane and nitrous oxide emissions; 6) increasing energy efficiency; 7) stopping deforestation, net of reforestation; 8) increasing carbon uptake through photosynthetic production of biomass; and 9) developing and scaling up new technology, including carbon capture technology.

The 2016 Paris Agreement seeks to involve all nations in a coordinated effort to limit global warming to 2°C or less relative to pre-industrial temperatures by the end of the 21st century. The agreement also aims to strengthen the ability of all countries to adapt to the effects of climate change. As of 2019, 195 countries have signed the Agreement and 184 have ratified it. The US initially signed the Agreement but withdrew in 2018. As of the end of 2018, nearly all signatories were falling short of their goals. From 2005 to 2017, global CO_2 emissions increased by ~24%. China and India increased CO_2 emissions while the US reduced CO_2 emissions by 15%. Carbon emissions can be significantly reduced if population growth is controlled, energy efficiency improves, and energy consumption is limited to 750 to 800 EJ in 2050.

Technology development will be needed to mitigate climate change. Relying solely on new technology to mitigate increased carbon emissions is a risky strategy. Economic incentives and disincentives can be used to promote lowering carbon emissions, such as a carbon tax or cap-and-trade carbon allowances. Short-term economic and political interests often prevail over the long-term interest of mitigating climate change. Without the leadership of China, the US, and India, meaningful progress is very unlikely.

Natural resources are either renewable or non-renewable. Non-renewable and renewable resources may become scarce. Non-renewable resources may be depleted. Conservation methods include recycling, reducing waste or unnecessary loss, preventing contamination or pollution, and developing alternative resources, including new technologies. Government regulation and changing human behavior are methods that can be used to alter the rate of consumption of natural resources and especially non-renewable resources.

The world will need to produce 20 to 22 ZJ of energy between now and 2050 and 70 to 80 ZJ of energy by the end of the century. Global primary energy consumption in 2018 was 0.58 ZJ. Global primary energy consumption from 2007 to 2017 grew at 1.5% per year. Global primary energy consumption from 2017 to 2018 grew 2.9%. Current recoverable reserves of oil and natural gas are about 11 ZJ for oil and 13 ZJ for natural gas. From 2007 to 2017, global demand for oil grew 1% per year and for natural gas grew 2.2% per year. The 2018 global proven reserve to production ratio for both oil and natural gas are about 50 years, based on current consumption rates. Even if global total oil and natural gas reserves are twice as large as 2018 proven global reserves, future increased production will deplete total oil and natural gas reserves by the end of the century.

In the US, more than 15% of oil and about 10% of natural gas production is used to produce petrochemicals. Examples of end products produced from petrochemicals include pharmaceuticals, cleaning products, rubber, cosmetics, insulation, lubricants, plastics, fabrics, animal feed, and paint. Demand for petrochemicals is expected to increase substantially during the remainder of this century. Petrochemicals are likely to account for more than a third of the growth in world oil demand by 2030, and nearly half the growth by 2050. Unless new technologies are developed to produce petrochemicals or they are produced from coal or other sources, conservation of global oil and natural gas resources will be needed to supply future generations with the raw materials for petrochemical production.

Global coal reserves are large, over 1 trillion tonnes. If global carbon emissions are to be stabilized and eventually reduced, use of coal for generating electricity must be reduced dramatically, unless highly effective and affordable flue gas carbon capture technology is developed and deployed on a large scale. Global reserves of coal will not be depleted this century. Coal reserves will likely be depleted in some geographic regions, such as China and possibly India. The US will continue to have sizable reserves long into the future.

Power generation from biomass and hydroelectric sources are unlikely to increase and may decrease because of climate change. In 2018 nuclear power plants supplied only about 11% of global electricity demand and about 1.5% of global total energy demand. Existing, affordable uranium ore reserves are about 8.4 million tonnes. Global uranium consumption is about 86,000 tonnes per year, resulting in a reserve to production ratio in 2018 of about 100 years. Economical U_{235} will likely be scarce by the end of the 21st century.

Assuming continued improvement in agricultural productivity and reduction in food waste, the amount of arable land on Earth should be able to support as many as 10-14 billion people. Global warming, droughts, desertification, urbanization, and pollution could reduce this number below 10 billion unless steps are taken to conserve arable land and use it optimally. Asia and Africa have the largest amount of land but have much less arable land per capita than Europe, Australia, North America, and South America. Africa does not have adequate arable land to meet future needs. African countries need to increase agricultural investment and crop productivity, decrease food waste, irrigate and fertilize pastureland, and import grain and cereals from regions of the world with excess production capacity. If climate change should reduce crop productivity, especially in Nigeria, Sudan, Ethiopia, and Tanzania, it is unlikely that sufficient arable land will be available to support the projected population growth in Africa.

It is crucial to protect available soil resources from degradation. There are a number of steps that can be taken: 1) develop more heat, draught, and disease resistant crops; 2) optimize use of fertilizers and irrigation; 3) prevent pollution or depletion of soil nutrients; 4) promote land reform and use of information technology; and 5) improve infrastructure to reduce food waste.

Fresh water is a finite but renewable resource. Europe, Africa, and South America have the largest reserves of fresh groundwater. Most of the rainfall occurs in South America, Central Africa, and South Asia. In the developed world most freshwater is used to generate electricity while in the developing world it is used for irrigation. Some 50 countries, with roughly a third of the world's population, experience water stress, and 17 of these extract more water annually than is recharged through their natural water cycles. By 2050 India and South Asia will experience water scarcity and the Middle East and North Africa will experience absolute water scarcity. Sub-Saharan Africa, Southwest North America, and Northwest China will experience water stress. The rest of the world would appear to have adequate water resources, unless climate change produces more extreme conditions than currently expected. There are a number of steps that can be taken to conserve water resource: 1) improve irrigation methods; 2) increase water catchment and storage; 3) build desalination plants; 4) recycle wastewater; 5) prevent water pollution; 6) improve sewage treatment and removal; 7) reduce use of water for electricity generation; 8) improve piping and water infrastructure; and 9) employ information technology to build "smart" water systems.

CHAPTER 7

Changing Course

"Problems are inevitable. All problems have a solution."
—David Deutch

FOR MANY DECADES, POPULATION GROWTH, climate change, and natural resource depletion have been identified as potential threats to mankind. In the 18th century Thomas Malthus developed a theory that human population growth would be exponential while growth in the food supply would be arithmetic. He concluded that these differential growth rates would lead to famine and economic decline.[125] While his food supply projections were not correct, his work led others to postulate a causal relationship between exponential population growth and depletion of non-renewable natural resources. In the 19th century Svante Arrhenius described the effects of atmospheric carbonic acid concentrations on global surface temperatures.[126] His initial work explained periods of global glaciation but was later applied more specifically to describing the effect of carbon emissions on global surface temperatures. In the 20th century M. King Hubbert developed methods to forecast depletion of non-renewable natural resources. [127]

In 1945 the United Nations was chartered to promote peace and intergovernmental cooperation. In 1972 the UN Environment Programme (UNEP) was organized to guide and coordinate environmental activities within the UN and promote the sustainable use of the world's natural resources. In 1974 The UN Population Action Plan was adopted and in 1988 the World Meteorological Organization and the UN organized the Intergovernmental Panel on Climate Change (IPCC). Along with many national

---

125 T. Malthus, *An Essay on the Principle of Population, as It Affects the Future Improvement of Society* (London: J. Johnson in St Paul's Churchyard, 1798).

126 S. Arrhenius, "On the Influence of Carbonic Acid in the Air upon the Temperature of the Ground,, " *The London, Edinburgh, and Dublin Philosophical Magazine and Journal of Science*: 41(251) (1896): 237-276, doi:10.1080/14786449608620846.

127 M. King Hubbert, "Energy Resources," National Academy of Sciences, Publication 1000-D (1962): 57.

and regional governments and numerous non-governmental organizations, plans were developed and implemented to reduce population growth and carbon emissions and conserve natural resources. Considerable financial, human, and intellectual capital has been spent over the past 50 years to mitigate these threats. Unfortunately these efforts have fallen short. This is not to say progress wasn't made or that these programs failed. It is to say that the magnitude of the progress has been insufficient.

There are many reasons progress has been insufficient. Public acceptance of the threat posed by population growth, climate change, and natural resource depletion has been insufficient to cause many national governments to consistently implement and maintain effective mitigation strategies. Public acceptance of these threats is increasing but it remains to be seen if it is now sufficient to cause governments to change policies and implement programs that are necessary to change course.

We know what the problems are and why they are occurring. We have solutions that work, and we have ongoing research and development programs to develop even better technologies and solutions. We have studied prior failures and know what doesn't work. We have the financial and human resources for implementation. We know how to organize and implement complex programs. We know how to measure progress. We know when these actions should be taken and when to expect results. However, resistance from certain political groups, influential individuals, governments, trade organizations, and businesses that see proposed mitigation efforts as being incorrect, too expensive, immoral, corrupt, or otherwise antithetical to their special interests has impeded progress. In addition, entrenched cultural and religious practices and beliefs have prevented or undermined programs that might have otherwise been effective in controlling population growth.

The majority of the world's population has no understanding of the perfect threat, why it exists, and least of all what to do about it. Most people on Earth live at or just above a subsistence level. At this level, people are concerned with their immediate safety and obtaining the essentials of life. Personal needs, deeply ingrained cultural and religious beliefs, and traditional rivalries drive behavior, not slowly developing, long-term threats that can't be seen or easily understood. In many regions of the world, a lack of stable government, corruption, and misappropriation of scarce resources has been sufficient to undermine progress or prevent otherwise effective programs from being implemented. A lack of access to adequate financial and human capital to implement mitigation strategies is a serious problem, especially in the developing world.

While the OECD has periodically committed to help fund mitigation programs, the level of support is often inconsistent or insufficient. The impediment caused by long-standing political animosity and conflict between nations, cultural groups, or religions cannot be underestimated. These "tribal" conflicts lead to suspicion, lack of trust, and unwillingness to compromise or find common ground. If agreements are forged, they are often voluntary and lack an effective mechanism to hold governments, political leaders, businesses, and institutions accountable for failing to meet goals.

We live on a small planet and that planet is getting much more crowded. Population growth and economic development are the primary causes of environmental damage, including climate change and natural resource depletion. The good news is that many developed countries and regions of the world have already reduced total fertility rates to replacement levels or below. However, the population growth rate in Africa is greater than 2% per year and the total fertility rate in 50 of 137 developing countries is greater than 3.5 live births per woman of childbearing potential.

Population momentum in the developing world will continue to drive population increases even if the total fertility rate is reduced substantially. There are 10 African and 4 Middle Eastern countries that will account for more than 90% of the projected population growth through 2100. Most of these countries have unstable, authoritarian governments and conservative cultural or religious values. The population growth rate among Muslims is 1.5% per year compared to 0.7% for all other religions and agnostics combined. Future population growth can be addressed with programs and policies that focus on 10-20 countries. Population momentum will be very difficult to overcome but there are examples of nations that have been successful in bending the curve over 20-30 years. It requires stabilizing governments, enlisting the active support of religious leaders, organizing effective educational and health care programs, recruiting experienced teams, and providing the necessary financing, but it can be done.

Many barriers have impeded progress in reducing the rate of climate change and natural resource depletion and adapting to their consequences. In 2018 the president of the United States asserted that climate change is a hoax and withdrew the US from the Paris Climate Agreement because it was unfair to the US. This Agreement allows China and India to claim status as developing countries, despite their position as the second and fifth largest economies on Earth, and to increase their carbon emissions from 2016 to 2030 by almost 50% and still meet their Nationally Determined Contributions (NDCs).

All NDCs are self-determined and voluntary, and very few countries are even marginally meeting their goals. As a result, the Paris Agreement will not limit global warming to 2°C or less relative to pre-industrial levels this century. However, withdrawal of the US from the Agreement and the tacit approval by the international community of substantial increases in carbon emissions from China and India during the next ten years will make it impossible to prevent a rise in global surface temperatures of less than 3.0°C this century.

Investment in oil and natural gas exploration and development is greater than $400 billion per year while investment in solar and wind technology is only about $300 billion per year. In the last 20 years about 370 million acres of arable land have been added through deforestation and irrigation and about 360 million acres of arable land have been lost due to desertification, urbanization, pollution or sea level rise. These failures are due to poor judgment and miscalculation, and the large political and financial cost associated with implementing effective mitigation strategies. They serve to illustrate how ill equipped humans are to address risks that are remote, develop slowly, and are impersonal, even if they have profound consequences for future generations.

An integrated, comprehensive, international plan that addresses population growth, climate change, and natural resource depletion and holds both political and programmatic leadership accountable for performance is missing. Without reducing population growth it is unlikely that global warming and natural resource depletion can be mitigated. However, reducing population growth alone is not sufficient to ensure global warming and natural resource depletion will be reduced. All three threats need to be effectively addressed together.

Adequate human and financial capital must be provided. Economists have evaluated the relationship between the economic cost of global warming and the degree of global warming.[128] (1) The Environmental Impact and Sustainability Applied General Equilibrium (ENVISAGE) model was developed and is maintained by The World Bank. The Dynamic Integrated Climate-Economy (DICE) model was developed by William Nordhaus, a Yale University Professor and Nobel Prize winning economist; and the Climate and Regional Economics of Development (CRED) model was developed by the Stockholm Environment Institute. These models give damage estimates that are similar at low to moderate levels of global warming, but diverge at higher levels, reflecting their differing assumptions. (Figure 7.1)

Global warming has already reached 1.1°C above pre-industrial levels and will likely reach 2.9 ± 0.3°C by 2050. The change from 2050 to 2100 depends on the degree to which mankind is able to mitigate carbon emissions. If no action is taken, global mean surface temperatures will likely increase by 4.8 ± 0.4°C relative to pre-industrial temperatures by the end of the century with further warming expected in the 22nd century. The economic models suggest damages would range from 2% to 4.5% of GDP at a 3°C increase in global mean surface temperatures up to 4% to 14% or more at a 5°C increase. The world currently spends about 2.1% of GDP on military spending and 10% on health care. The world will need to divert at least 2.5-3.0% of global GDP ($2.2 to $2.6 trillion) per year to effectively mitigate the perfect threat.

---

128   J. Harris, B. Roach, A-M Codur, "The Economics of Global Climate Change" (2017), http://ase.tufts.edu/gdae.

To succeed, at least as much money needs to be spent annually as is devoted to military spending and possibly as much as is devoted to healthcare. The human and financial cost of failing to make this investment is enormous. To create adequate financial resources, revenues will need to be raised and spending in non-essential areas will need to be reduced. The global financial impact will be substantial, and the economic effects will be felt at all levels of society.

Figure 7.1. Relationship of changes in global mean temperature with damages to the world economy[129]

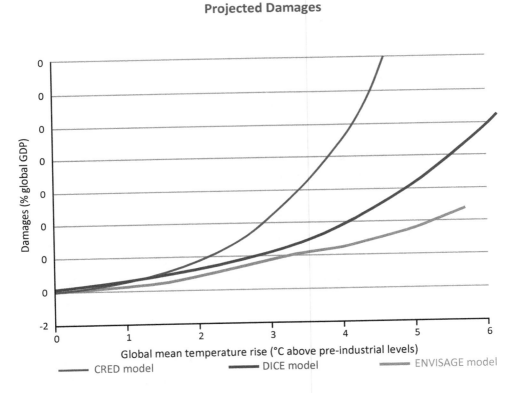

Coordinated effective leadership from China, India, and the US is needed. These countries have very different cultures, political systems, and economies. In 2018, China, India, and the US represented 43% of world GDP, 41% of world population, and more than 50% of world carbon emissions. They are political and economic rivals. Their vulnerability to population growth, climate change, and natural resource depletion are different. India and China are vulnerable and face major threats with serious political, social, and economic consequences. North America is far better positioned but still faces major challenges. Europe, Japan, and Russia also have to be a part of the solution since Europe and Japan

---

129 Harris, Roach, and Codur, *"The Economics of Global Climate Change."*

are important world economies and Russia has abundant natural resources. The United Nations and other countries may play key roles and serve as valued partners.

Failure to receive the full support and firm commitment of the US, China, and India will lead to failure. Existing commitments from these countries are inadequate and inconsistent. Religious leaders of the major Christian, Hindu, and Islamic sects must actively and vigorously support any international plan. There are significant cultural and traditional beliefs that serve as major barriers to achieving success. Effectively navigating through the social, political, and economic obstacles standing in the way of progress is and will likely continue to be extremely difficult.

Creativity and critical thinking are essential to the generation of new knowledge and are fundamental for problem solving. Mankind must continue to generate new knowledge and rely on adaptability and creativity to ultimately solve future problems. We must be open to new ways of thinking, new technologies, new social structures, and new political institutions since this is where solutions are likely to emerge.

In 2100 the world population could be 8.5 billion or less. Effective national and international programs could exist to promote controlled migration, education and training, cultural assimilation and employment of migrants to meet the economic needs of countries with a declining population. Atmospheric carbon-dioxide equivalent concentrations could be stable at ~650 ppm, Global average surface temperature increases could be limited to a rise of ~3°C from pre-industrial levels. Technology could be developed and scaled up to capture and store or utilize carbon emissions. Deforestation could be stopped and reforestation increased. Pollution and contamination of natural resources, especially the oceans, fresh water sources, and arable land could be stopped. Effective re-cycling programs could be expanded. Measures to effectively adapt to the consequences of higher global mean surface temperatures could be widely adopted and implemented. The world's collective response to the threats posed by population growth, climate change, and natural resource depletion, including our use of existing technology or development of new technology, will largely determine mankind's fate in the 22nd century.

GLOSSARY

Language is essential to the effective transfer of knowledge. In general, the language used in this book follows common American usage that can be found in any standard, authoritative dictionary. Some technical or specialized terms are defined in the body of the text, although many of these definitions are repeated here so this glossary can serve as a common reference.

The International System of Units is the primary measurement system used. The kilogram and meter are the standard measures of weight and length, respectively. Temperature is measured on the Celsius scale, although the Fahrenheit scale is often noted in parentheses for clarity and to avoid the need for the reader to convert scales. Energy is measured in joules, power in watts, and volume in liters. Occasionally, the US customary system for weight, length, volume, and area is used in addition to the International System of Units for illustration and/or clarity.

Acronyms and Abbreviations

CO_2: Carbon dioxide

ENSO: El Niño Southern Oscillation

FAO: Food and Agriculture Organization of the United Nations

GMST: Global mean surface temperature

GDP: Gross Domestic Product

GISS: Goddard Institute for Space Studies

GtC: Gigatonne carbon

GWP: Global warming potential

Ha: Hectare

IGCC: Integrated gasification combined cycle

IPCC: Intergovernmental Panel on Climate Change

Kg.: Kilogram

Km.: Kilometer

Km^3: Cubic kilometer

L-OTI: Land-Ocean temperature index

M^2: Square meter

MJO: Madden-Julian Oscillation

NASA: National Aeronautics and Space Administration

NDC: Nationally determined contribution

NOAA: National Oceanic and Atmospheric Administration

OECD: Organization for Economic Co-operation and Development

S: Second

TFR: Total Fertility Rate

TWS: Terrestrial water supply

UN: United Nations

UNFCCC: United Nations Framework Convention on Climate Change

US: United States of America

UV: Ultraviolet radiation

Definitions
General terms

Anthropogenic: caused or produced by human beings.

Developed Country: A developed country has an economy with large industrial and service sectors, high per capita GDP, and advanced transportation, financial, and energy infrastructure.

Developing Country: A developing country has a low per capita GDP, basic infrastructure, and an economy that is largely based on agriculture or fishing.

Economic Growth: The rate of increase in real GDP. Real GDP is the inflation adjusted total value of goods and services produced by an economy over a specified time period. GDP is calculated by summing the amount of money spent by government, by individuals, and by businesses, plus the balance of trade, for an economy.

Efficient: Producing desired results without waste. The ratio of the useful work obtained to the total work expended is high.

Error rate: The ratio of the number of errors in a data set to the number of data points in the same data set.

Fossil fuel: A fossil fuel is a combustible organic compound or compounds formed over millions of years from the decomposition of living organisms by the heat and pressure of the Earth's mantle. Oil, natural gas, and coal are the principal fossil fuels.

Likely: Greater than an 80% chance (8:10) an event occurred or that a future event will occur. *Very likely* is >95% chance.

Reliable: Reliable means that an effect will likely be reproduced if evaluated under the same or similar conditions in the future. Reliability is a measure of consistency. Reliable results are reproducible.

Safe: Secure from the threat of danger, not risky, not likely to cause significant harm or loss.

Unlikely: Less than a 20% chance (1:5) an event occurred or that a future event will occur. *Very unlikely* is less than a 5% chance.

Valid: Valid means the result obtained is correct, can be generalized, and is reliable.

Geography

Asia: Asia is the largest continent on Earth and includes the following regions:

a. Southern Asia: Bangladesh, India, Afghanistan, Pakistan, Bhutan, Nepal, and the Maldives.
b. Eastern Asia: China, Hong Kong, Macau, Taiwan, Japan, Mongolia, North Korea, and South Korea
c. Western Asia: Afghanistan, Azerbaijan, Bahrain, Gaza Strip, Iran, Iraq, Israel, Jordan, Kuwait, Lebanon, Oman, Qatar, Saudi Arabia, Syrian Arab Republic, Turkey, United Arab Emirates, West Bank and Yemen

    d.   Southeastern Asia: Brunei, Myanmar, Cambodia, Timor-Leste, Indonesia, Laos, Malaysia, the Philippines, Singapore, Thailand and Vietnam.

    e.   Central Asia: Kazakhstan, Kyrgyzstan, Tajikistan, Turkmenistan, and Uzbekistan

Africa: Africa includes Egypt and all of the other countries on the continent of Africa.

Europe: Europe includes the countries of the European Union plus Albania, Andorra, Armenia, Azerbaijan, Belarus, Bosnia, Georgia, Iceland, Kosovo, Liechtenstein, Moldova, Monaco, Montenegro, North Macedonia, Norway, Russia, San Marino, Serbia, Switzerland, Ukraine, and Vatican City. It does not include Turkey.

European Union: The European Union includes Austria, Belgium, Bulgaria, Croatia, Cyprus, Czechia, Denmark, Estonia, Finland, France, Germany, Hungary, Ireland, Italy, Latvia, Lithuania, Luxembourg, Malta, Netherlands, Poland, Portugal, Romania, Slovakia, Slovenia, Spain, Sweden, and The United Kingdom.

Latin America: Latin America includes Mexico and all of the countries of Central and South America.

Middle East: The Middle East consists of the countries of Western Asia with the exclusion of Afghanistan and Azerbaijan and the addition of Egypt and Turkey.

North America: North America is the United States of America and Canada. Mexico is not included.

Oceania: Australia, New Guinea, New Zealand, Tasmania, and the Pacific Islands of Melanesia and Micronesia. It does not include Indonesia or the Philippines.

Population Growth

Birth Rate: The number of births in a particular population per unit of time usually expressed as the number of births per 1000 individuals per year.

Census: A survey of a population with the aim of a complete enumeration and characterization of the population.

City: A city is a large, concentrated human population. A city usually has political boundaries and political leadership and may also have an adjacent area that is functionally related to and economically dependent on the city but not a part of the political boundaries.

Death Rate: The number of deaths in a particular population per unit of time usually expressed as the number of deaths per 1000 individuals per year.

Demography: Demography is the study of human populations. Demographers use a variety of methods to measure the size, structure, and distribution of human populations and any spatial or temporal changes that may occur. Demographers use direct and indirect methods to measure and characterize populations.

Estrus Cycle: The estrus cycle is the reproductive cycle for most female mammals. The cycle is determined by periodic changes in the production of certain hormones that induce ovulation and changes in the female reproductive tract that promote fertilization. A female mammal in estrus can become pregnant and exhibits behavior indicating she is receptive to mating.

Exponential growth: Exponential growth is a specific way something can change over time. In this case the change is an instantaneous increase that is proportional to the starting value. For example, an instantaneous doubling of the starting value

Genus: A genus is a rank or grouping used in the classification of biological organisms. The order of biological ranks is Domain, Kingdom, Phylum, Class, Order, Family, Genus, and Species. All members of a genus have similar (but not identical) genotypes and have common phenotypic characteristics that can be used by taxonomists for classification.

Population Growth Rate: the rate at which the number of individuals in a population increases in a given time period, expressed as a fraction or percentage of the initial population.

Population Momentum: Population momentum accounts for the effect of the number of women of childbearing potential in a population on population growth. Population momentum is the ratio of the size of a population at equilibrium, when the fertility rate and the replacement rate are equal and the number of women of childbearing potential is not changing, to the current size of the population. Population momentum causes a population to continue to grow even after the fertility rate and replacement rate are equal if the number of women of childbearing potential is increasing.

Population Pyramid: A population pyramid depicts the percent of the population by gender in each 5-year age increment from birth to 100+ years.

Species: A species is a rank or grouping used in the classification of biological organisms. The order of biological ranks is Domain, Kingdom, Phylum, Class, Order, Family, Genus, and Species. Taxonomists can define members of a species in several ways. *Homo sapiens* (modern humans) are all members of the genus *Homo,* share certain common genetic sequences, have similar morphology, and can mate and produce fertile offspring.

Total Fertility Rate: The average number of children that would be born to a hypothetical typical woman over her lifetime if she was to survive from birth to the end of her reproductive life.

Refugee: A refugee is someone who must leave a country because of an imminent threat. Refugees are protected by international agreements.

Replacement Rate: the total fertility rate needed to sustain current population levels. The replacement rate is typically above 2.0, the theoretical replacement rate, because of female mortality during reproductive years.

Transnational migrant: A transnational migrant is someone who decides to move from one country to another in search of a better life. Migrants are not protected by international agreements, although in 2018 the Marrakesh Compact was endorsed by the United Nations to provide non-binding international standards for the treatment of transnational migrants.

Under-employment: Underemployment occurs when a person is employed but their productivity is below the level that could be achieved during the time the worker is available to work. It does not count voluntary (or willing) part-time employment. It does count involuntary (or unwilling) part-time employment and under-utilization of the experience and abilities of a worker. Underemployment contributes to low productivity in an economy.

Unemployment: Unemployment occurs when a person is able to work, is employable, and is actively looking for a job but cannot obtain one. This circumstance creates a state of involuntary idleness. It does not count people who are disabled or cannot work, and it does not count voluntary idleness. It does not count people who are retired, under legal working age, disabled, in school, or who for personal reasons are able to work but are not actively seeking a job.

Urbanization: Urbanization is an increase in the population of cities.

Climate Change

Aerosol: An aerosol is a suspension of fine liquid or solid particles in a gas, such as air. The particles are usually less than 1μm in size.

Albedo: Albedo is the proportion of incident light or radiation that is reflected by a surface, typically that of a planet or moon. The degree to which the surface and atmosphere of the Earth are reflective is called the Earth's albedo. A surface that is a perfect reflector has an albedo of 1.0.

Carbon cycle: The carbon cycle is the process by which carbon is exchanged among the different reservoirs of carbon on Earth, such as the biosphere, atmosphere, rocks, and the oceans. There is a fast component and a slow component to the carbon cycle. The fast component primarily describes the flux of carbon between the atmosphere and biosphere. The slow component describes the movement of inorganic carbon between the hydrosphere and geosphere.

Carbon dioxide: At room temperature, carbon dioxide is a colorless, odorless gas consisting of one carbon atom and two oxygen atoms (CO_2) covalently bound. Carbon dioxide is produced by burning organic compounds or by respiration and is consumed by photosynthesis. When dissolved in water it produces carbonic acid.

Climate: Climate is the long-term pattern of temperature, precipitation, humidity, barometric pressure, cloud cover, and wind. The time period used to evaluate climatic conditions and determine if a change has occurred is usually 30 years or more. Generally, the climate of a terrestrial region is related to its latitude, altitude, terrain, land cover, and proximity to oceans and their currents.

Climatic Zone: A Climatic zone is any of the geographical zones loosely divided according to prevailing climate and latitude.

Convection: Convection is a type of heat transfer that occurs in liquids or gases when molecules move (flows) in response to differences in density (temperature) within the liquid or gas.

Desertification: The process by which fertile land is converted into a desert due to changes in temperature, draught, deforestation, or contamination.

Energy budget: The energy cycle describes the flux of energy derived from the Sun and Earth's core between the reservoirs of energy on Earth, or back into space. The principle terrestrial reservoirs are the hydrosphere (e.g., the oceans), atmosphere (including clouds), geosphere, and biosphere.

Flux: The rate of flow across, or incident on, an area. Flux usually refers to radiant energy, magnetism, particles, or fluids.

Geothermal flux: Geothermal flux is the amount of heat moving steadily outward from the interior of the Earth through a geographic area over a specified period of time.

Global insolation: Global insolation is the amount of solar radiation that reaches the Earth's surface; measured in $MJ/m^2/day$.

Greenhouse gases: Greenhouse gases are gases in the atmosphere that absorb infrared radiation and reradiate much of it back toward the Earth's surface. These gases include water vapor, carbon dioxide, methane, nitrous oxide, and chloroflurocarbons.

Heat capacity: The heat capacity of a substance is the amount of energy required to heat one unit of the substance one degree Kelvin (or one degree Celsius) under specified pressure conditions. The units of heat capacity can be expressed as mass (e.g., grams), volume (e.g., cubic centimeters), or other standard quantities (e.g., moles).

Infrared radiation: Infrared radiation (IR) is a region of the electromagnetic radiation spectrum with wavelengths ranging from about 700 nanometers (nm) to 1 millimeter (mm). Infrared waves are longer than those of visible light, but shorter than those of radio waves. Humans perceive infrared radiation as heat.

Net energy balance: The ratio of energy output relative to the energy input needed to produce the energy, usually used in the context of a fuel. A value greater than one indicates that more energy was obtained from the process than the energy needed to produce it.

Permafrost: Permafrost is defined as subsurface soil or rock that remains at or below 0°C for at least two consecutive years. Most of the permafrost existing today formed during glacial periods and has persisted in Polar Regions, especially in the northern hemisphere.

Photosynthesis: Photosynthesis is a biochemical process that is essential to life on Earth. Photosynthesis is carried out primarily by photoautotrophs such as plants and algae. These organisms contain pigments such as chlorophyll that can absorb solar energy and use it to split water into hydrogen and oxygen. Hydrogen is then combined with carbon dioxide in a biochemical process called the Calvin cycle to produce carbohydrates that can be used for other cellular anabolic pathways.

Photosynthetically active radiation (PAR): Photosynthetically active radiation (PAR) is the spectral range of solar radiation from 400 to 700 nanometers that photosynthetic organisms use in the process of photosynthesis.

Radiative Forcing: Radiative forcing is the difference between total solar irradiance and the amount of radiation emitted by the Earth back into space. Positive radiative forcing means the Earth receives more energy from the sun than it radiates back into space.

Respiration: Respiration is the physiological process through which oxygen in the environment is exchanged with carbon dioxide and water in an organism.

Solar cycle: The solar cycle or solar magnetic activity cycle is a periodic 11-year change in the sun's activity (including changes in the amount of solar radiation and ejection of solar material) and appearance (changes in the number and size of sunspots, flares, and other manifestations).

Solar Radiation/Irradiance: Solar radiation or irradiance is the power (watts/m²) emitted by the sun from nuclear fusion reactions that create broad-spectrum electromagnetic energy. The solar power reaching the top of the Earth's atmosphere is called Total Solar

Irradiance. The magnitude of TSI is dependent on the distance of the Earth from the sun and the solar cycle.

Stratosphere: The stratosphere is the layer of the atmosphere above the troposphere. It extends from ~10 km to ~50-60 km above the surface of the Earth. Temperatures are stratified in the stratosphere, ranging from -51°C in the lower regions to -15°C at the top of the stratosphere. Temperature increases with increasing altitude due to the absorption of solar radiation by the ozone layer.

Surface temperature anomaly: Surface temperature anomaly refers to the difference between a reference temperature and the actual temperature for a given region. The *global mean surface temperature anomaly* is determined from a data set consisting of monthly average surface temperature anomalies on a 5° x 5° grid across the land and ocean surfaces of the Earth. The reference period is a 30-year period containing the most recent decade, currently 1981-2010.

Thermal convection: Thermal convection is the transfer of heat from one place to another by the movement of fluids.

Transpiration: Transpiration refers to the release of water vapor through the stomata of leaves.

Troposphere: The troposphere is the lowest level of the Earth's atmosphere. It extends from the Earth's surface to a height of ~10 km (Range: 7-20 km). The troposphere contains almost all of the moisture in the atmosphere and almost all weather occurs in this layer. Air temperature decreases with increasing altitude in the troposphere. When the temperature gradient decreases to less than 2°C/km, the troposphere ends, and the stratosphere begins (tropopause).

Ultraviolet radiation: Ultraviolet (UV) radiation is electromagnetic radiation or light having a wavelength greater than 10 nm but less than 400 nm. Ultraviolet radiation has a wavelength longer than that of x-rays but shorter than that of visible light. UV radiation has both positive and negative effects on human health. The spectrum is divided into UV-A, UV-B, and UV-C. UV-B and UV-C are absorbed by the ozone layer in the stratosphere.

Weather: Weather describes the short-term pattern of temperature, precipitation, humidity, barometric pressure, cloud cover, and wind. Changes in the weather do not necessarily indicate that a change has occurred in a region's climate.

Natural Resource Depletion

Actinium: Actinium is a chemical element that is radioactive and can be found in uranium ores.

Breeder reactor: A breeder reactor is a nuclear reactor that produces more fissile material than it consumes. Plutonium is usually produced from U_{238} in these reactors.

Capacity factor: The capacity factor is the ratio of actual output divided by the nameplate capacity. Measuring or estimating the actual energy produced in a year and dividing by the nameplate capacity for a year determines the "capacity factor" for a power plant.

Energy: In physics, energy is the capacity for doing work. Energy can exist in many forms, including potential, chemical, thermal, mechanical, and nuclear. The first law of thermodynamics dictates that energy cannot be created or destroyed; it can only be converted from one kind to another.

Energy Efficiency: Energy efficiency is the ratio of energy input to energy output in a device, process, or economy. In economic terms it is the amount of energy produced per unit of GDP, often expressed as MJ/$GDP.

Fuel cell: A fuel cell is a device that converts chemical energy into electrical energy through a process that requires the use of specific fuels and catalysts. Fuel cells produce a direct electrical current continuously as long as the fuel is provided to functional catalysts. Typically the fuels are hydrogen and air (oxygen), and the emissions are oxygen and water. There are no carbon emissions from a fuel cell.

Geothermal energy: Geothermal energy is energy that is derived from the heat of the Earth's mantle. This energy comes from the Earth's interior and is independent of solar radiation.

Heavy oil: Heavy oil is highly viscous, dense oil that cannot easily flow in production.

Hydrocarbon: A hydrocarbon is an organic compound consisting entirely of hydrogen and carbon.

Levelized cost of electricity: The levelized cost of electricity is the ratio of the total cost of ownership to the total actual electricity production over the lifetime of a device, facility, or type of energy. The costs are fully loaded so they include the cost of construction or production and financing, as well as operating and maintenance costs.

Nameplate capacity: Nameplate capacity is the amount of output produced at peak operation of a facility or device under ideal conditions. As applied to power plants, this term refers to the peak output of power under ideal conditions. The amount of energy produced by a power plant at nameplate capacity is equal to the nameplate capacity times the amount of time of operation. For example a 100 MW power plant operating under ideal conditions for one year would produce 100 MW X 8,760 hours per year or 876,000 megawatt-hours of energy.

Overnight construction costs: The cost of construction as if no interest was incurred during construction.

Photovoltaic Cells: Photovoltaic cells are devices for converting electromagnetic radiation into electrical energy using semiconductors, such as silicon, that produce the photovoltaic effect upon exposure to sunlight. The photovoltaic effect frees electrons in a semiconductor to produce an electrical current.

Power: Power is the amount of energy used (transferred) or work done per unit of time.

Renewable energy: Renewable energy is energy produced from resources such as biomass, waterfalls, wind, sunlight, rain, tides, waves, and geothermal energy.

UNITS OF MEASUREMENT

Acre: An acre is not a metric unit. It is 43,560 square feet.

Btu: A British thermal unit (Btu) is a unit of energy. There are about 1055 joules in a Btu.

C: Celsius is a temperature scale. °C is the unit of this scale. The freezing point for water is 0°C and the boiling point is 100°C. One (1) degree Celsius is equal to 1.8°Farenheit.

Gigatonne: A gigatonne is a billion (10^9) tonnes and is roughly equivalent to the weight of four Empire State buildings.

Hectare: A hectare is 10,000 square meters and is equivalent to 2.47 acres.

Joule: The standard international unit of energy. A joule is the energy transferred to an object by attempting to move it one meter against a force of one Newton.

    a. *Megajoule*: A megajoule (MJ) is 10^6 joules (or a million joules).

    b. *Exajoule*: An exajoule (EJ) is 10^{18} joules (or a billion, billion joules). One exajoule is equal to 277.8 TWh.

    c. Zettajoule: A zettajoule (ZJ) is 10^{21} joules (or a billion, trillion joules) and is equivalent to the energy in about 164 billion barrels of oil.

Kelvin: The Kelvin scale is the primary measure of temperature in the International System of Units. The scale begins at absolute zero. The units of the Kelvin scale are the same magnitude of temperature as the units of the Celsius scale so 1K=1°C=1.8°F.

Kilogram: A kilogram is a unit of weight. It is equivalent to 2.2 pounds.

Kilometer: A kilometer is a unit of distance or length. It is equivalent to 0.62 miles or 3,280.84 feet.

Newton: The standard international unit of force. 1 newton of force is the force required to accelerate an object with a mass of 1 kilogram 1 meter per second per second. A force of one Newton is roughly equivalent to the force exerted by holding an apple in your hand at sea level. The units are kg•m•sec^{-2}.

Parts per million (ppm): Parts per million (ppm) refers to the concentration of a substance. It is a dimensionless number that is generally used to describe a low concentration of a substance. 1 ppm is equal to 1 mg/kg (1 milligram per kilogram) by weight or 1 mL/L (1 milliliter per liter) by volume. To put this number in perspective, 1 ppm is about 1 grain of sugar in 275 sugar cubes or 7 drops of ink in a bathtub of water.

Parts per billion (ppb): Parts per billion (ppb) refers to the concentration of a substance. It is a dimensionless number that is generally used to describe a very low concentration. 1 ppb is equal to 1 µg/kg (1 microgram per kilogram) by weight or 1 µL/L (1 microliter per liter) by volume. To put this number in perspective, 1 ppb is about 3 seconds out of a century.

Square kilometer: A square kilometer is a million (10^6) square meters, 100 hectares, or 247 acres.

Square mile: A square mile is not a metric unit. A square mile is 2.59 km^2 or 640 acres.

Ton: A ton is not a metric unit. A ton (or short ton) is 2,000 pounds or 0.907 tonnes.

Tonne: A tonne is 1,000 kilograms or 2,206.4 pounds.

Watt: A watt is a unit of power equal to one joule per second. The units are kg•m^2•s^{-3}.

    a. A *megawatt* (MW) is a million (10^6) watts.
    b. A *gigawatt* (GW) is a billion (10^9) watts. A one GW nuclear power plant operating at 90% capacity for one year provides 7.88 TWh of energy.
    c. A *terawatt* (TW) is a trillion (10^{12}) watts.

Watt-hour: A watt-hour is a unit of energy. The units are kg•m^2•s^{-2}.

    a. A *megawatt-hour (MWh)* is a million watt-hours or 1,000 kilowatt-hours (kWh). This amount of energy could supply an electric car for 3,600 miles or provide the energy needs of a typical American home for 1.1 months.
    b. A *gigawatt-hour (GWh)* is a billion watt-hours. This amount of energy could supply about 90 typical American homes for 1 year.
    c. A *terawatt-hour (TWh)* is a trillion watt-hours. A 1GW nuclear power plant operating at 90% capacity for 1 year produces about 7.88 TWh of energy. This amount of energy could serve about 700,000 American homes for one year. In 2019, the US produced 4,401.3 TWh of electricity.

FIGURES & TABLES WITH SOURCE INFORMATION

Figure 2.1: Estimated world population from 10,000 BCE to present.
Source: Our World in Data; https://ourworldindata.org.

Figure 2.2: Population Dynamics.
Source: Our World in Data; https://ourworldindata.org.

Figure 2.3: World Population Growth from 1950 to 2100.
Source: United Nations Department of Economic and Social Affairs, Population Division; https://population.un.org.

Figure 2.4: World birth rate and death rate from 1960 to 2020.
Source: United Nations Department of Economic and Social Affairs, Population Division; https://population.un.org.

Figure 2.5: The Population Pyramid for the World.
Source: PopulationPyramid.net; https://populationpyramid.net.

Figure 2.6: Projected size of the world population 2020 to 2100.
Source: United Nations Department of Economic and Social Affairs, Population Division; https://population.un.org.

Figure 2.7: Total Fertility Rates by Country, 2015-2020.
Source: Our World in Data; https://ourworldindata.org.

Figure 2.8: Average annual population growth rate by country.
Source: Our World in Data; https://ourworldindata.org.

Figure 2.9: 2018 population pyramids for China and Europe.
Source: PopulationPyramid.net; https://populationpyramid.net.

Figure 2.10: Population pyramids for Africa, India, Pakistan, Iraq, Egypt, and Indonesia.
Source: PopulationPyramid.net; https://populationpyramid.net.

Figure 2.11: Rural and Urban population from 1960 to present.
Source: Our World in Data; https://ourworldindata.org.

Figure 2.12: Urban population living in slums.
Source: Our World in Data; https://ourworldindata.org.

Figure 2.13: Total International Migrant Stock from 1960 to 2019.
Source: United Nations Department of Economic and Social Affairs, Population Division; https://population.un.org.

Figure 2.14: *Per capita* GDP from 1950 to 2017.
Source: Our World in Data; https://ourworldindata.org.

Figure 3.1: Solar cycles from 1880 to 2018.
Source: National Aeronautics and Space Administration (NASA), Goddard Institute for Space Studies (GISS); https://Climate.NASA.gov.

Figure 3.2: Milankovitch Cycles and Temperatures from the Vostok Ice Core.
Source: National Aeronautics and Space Administration (NASA), Goddard Institute for Space Studies (GISS); https://Climate.NASA.gov.

Figure 3.3: The Earth's albedo from March 2000 to 2013.
Source: National Aeronautics and Space Administration (NASA), Goddard Institute for Space Studies (GISS); https://Climate.NASA.gov.

Figure 3.4: The Earth's albedo anomaly from March 2000 to present.
Source: National Aeronautics and Space Administration (NASA), Goddard Institute for Space Studies (GISS); https://Climate.NASA.gov.

Figure 3.5: Solar radiation transmission through the Earth's stratosphere.
Source: National Oceanic and Atmospheric Administration (NOAA), Earth System Research Laboratories; https://www.esrl.noaa.gov.

Figure 3.6: Electromagnetic spectrum at the top of the Earth's atmosphere and at sea level.
Source: Vision Learning: https://visionlearning.com.

Figure 3.7: El Niño oscillation in 2016.
Source: Columbia University. https://blogs.ei.columbia.edu/2016/02/02/ el-nino- and-global-warming-whats-the-connection.

Figure 3.8: The Earth's energy budget.
Source: NASA: https://www.nasa.gov.

Figure 3.9: Radiative Forcing Components.
Source: Leland McInnes. https://commons.wikimedia.org.

Figure 3.10: The Earth's Fast Carbon Cycle.
Source: NASA. https://Earthobservatory.nasa.gov/features/CarbonCycle.

Figure 3.11: Adjusted global mean surface temperature anomaly (GMST).
Source: NASA. https://data.giss.nasa.gov.

Figure 3.12: Average L-OTI anomaly for 2014-2018.
Source: NASA. https://data.giss.nasa.gov.

Figure 3.13: Zonal mean L-OTI anomaly.
Source: NASA. https://data.giss.nasa.gov.

Figure 3.14: Global surface temperatures for the last 11,300 years.
Source: S. Bova, Y. Rosenthal, Z. Liu, S.P. Godad, and M. Yan, "Seasonal Origin of the Thermal Maxima at the Holocene and the Last Interglacial," *Nature* 589 (2021): 548-553.

Figure 3.15: GMST anomaly and total solar irradiance from 1880 to present.
Source: National Aeronautics and Space Administration (NASA), Goddard Institute for Space Studies (GISS); https://Climate.NASA.gov.

Figure 3.16: Air Temperature Anomalies over Land and Ocean since 1880.
Source: NASA. https://data.giss.nasa.gov.

Figure 3.17: Average global temperature anomalies related to volcanic activity.
Source: NASA. https://Earthobservatory.nasa.gov.

Figure 3.18: Atmospheric Carbon dioxide concentrations at Mauna Loa, Hawaii.
Source: NOAA Global Monitoring Laboratory. https://www.esrl.noaa.gov.

Figure 3.19: Changes in temperature, CO2, and dust over the last 800,000 years.
Source: Fabrice Lambert: https://commons.wikimedia.org/wiki/File:%22EDC_TempCO2Dust%22.svg.

Figure 3.20: CO2, N2O, CH4, and CFC concentrations from 1979 to 2018.
Source: NOAA Global Monitoring Laboratory. https://www.esrl.noaa.gov.

Figure 3.21: The Earth's energy budget.
Source: M. Wild, D. Folini, C. Schar, N. Loeb, E. G. Dutton, and G. Konig- Langlo, "The Global Energy Balance from a Surface Perspective," *Climate Dynamics* 40 (2013): 3107–3134, DOI 10.1007/s00382-012- 1569-8.

Figure 3.22: Representative Greenhouse Gas Concentration Pathways.
Source: van Vuuren, D.P., Edmonds, J., Kainuma, M. *et al.* The representative concentration pathways: an overview. *Climatic Change* **109,** 5 (2011). https://doi.org/10.1007/s10584-011-0148-z.

Figure 4.1: World energy consumption from 1800 to 2019 by fuel type.
Source: Our World in Data; https://ourworldindata.org.

Figure 4.2: US Energy Consumption by Source, 2019.
Source: US Energy Information Administration, *Monthly Energy Review* (2019), https://www.eia.gov/energyexplained/what-is-energy/sources-of-energy.php.

Figure 4.3: World land cover from the GLC-SHARE database.
Source: John Latham, Renato Cumani, Ilaria Rosati, and Mario Bloise, "Global Land Cover (GLC-SHARE)," Beta-Release 1.0 Database, Land and Water Division (2014), http://www.fao.org/uploads/media/glc-share- doc.pdf.

Figure 4.4: The Earth's water cycle.
Source: Carleton College. https://serc.carleton.edu/eslabs/drought/1b.html.

Figure 4.5: The world's major aquifers.
Source: R. Taylor, B. Scanlon, P. Döll et al., "Ground Water and Climate Change," *Nature Clim Change* 3 (2013): 322–329, https://doi.org/10.1038/nclimate1744.

Figure 4.6: Average annual precipitation.
Source: NASA. https://www.jpl.nasa.gov/edu/teach/activity/precipitation-towers-modeling-weather-data/.

Figure 4.7: The world's ground water resources.
Source: T. Gleeson, Y. Wada, M. F. P. Bierkens, and L. P. H. van Beek, "Water Balance of Global Aquifers Revealed by Groundwater Footprint," *Nature* 448 (2012): 197-200.

Figure 4.8: Fresh water availability per person per year.
Source: Our World in Data; https://ourworldindata.org.

Figure 6.1: Population pyramid for less developed and more developed regions.
Source: PopulationPyramid.net; https://populationpyramid.net.

Figure 6.2: Population pyramid for Pakistan.
Source: PopulationPyramid.net; https://populationpyramid.net.

Figure 7.1: Effect of GMST rise on global GDP.
Source: J. Harris, B. Roach, A-M Codur, "The Economics of Global Climate Change" (2017), http://ase.tufts.edu/gdae.

Tables

Table 2.1: Projection variants and their assumptions.
Source: United Nations Department of Economic and Social Affairs, Population Division; https://population.un.org.

Table 2.2: Mid-year 2019, 2050, and 2100 global population estimates.
Source: United Nations Department of Economic and Social Affairs, Population Division; https://population.un.org.

Table 2.3: Within Asia regional population growth estimates from 2019 to 2100.
Source: United Nations Department of Economic and Social Affairs, Population Division; https://population.un.org.

Table 3.1: Major factors determining global surface temperatures.
Source: C. Smith

Table 3.2: Atmospheric concentration, residence time and Global Warming Potential for common greenhouse gases.
Source: Massachusetts Institute of Technology. https://globalchange.mit.edu.

Table 3.3: Earth's carbon reservoirs.
Source: H. Riebeek and R. Simmon, "The Carbon Cycle" (2011), www.Earthobservatory. nasa.gov.

Table 3.4: Fast Carbon Cycle fluxes per year and the effect of the flux on atmospheric carbon concentrations.
Source: H. Riebeek and R. Simmon, "The Carbon Cycle" (2011), www.Earthobservatory. nasa.gov.

Table 3.5: Carbon dioxide emissions in 1970 and 2017 by region.
Source: Our World in Data: https://ourworldindata.org/co2-emissions.

Table 3.6: 2015 to 2017 fast carbon cycle sources and reservoirs.
Source: Global Carbon Project: https://www.globalcarbonproject.org/carbonbudget.

Table 3.7: Expected changes in the Earth's energy budget.
Source: C. Smith

Table 3.8: Predicted change in mean global surface temperature anomaly for RCP 4.5, 6.0, and 8.5.
Source: P. Brown and K. Caldeira, "Greater Future Global Warming Inferred from Earth's Recent Energy Budget," *Nature*: 552 (2017): 45-50.

Table 4.1: 2018 World energy consumption by energy source.
Source: British Petroleum Company, *BP Statistical Review of World Energy*, 69th edition (London: British Petroleum Co., 2020).

Table 4.2: Estimates of world conventional fossil fuel reserves.
Source: British Petroleum Company, *BP Statistical Review of World Energy*, 69th edition (London: British Petroleum Co., 2020).

Table 4.3: 2019 proven fossil fuel reserves at year-end.
Source: British Petroleum Company, *BP Statistical Review of World Energy*, 69th edition (London: British Petroleum Co., 2020).

Table 4.4: Undiscovered petroleum, natural gas, and natural gas liquids by geographic region reported by the US Geological Survey.
Source: US Geological Survey, "World Oil and Gas Resource Assessments" (2020), https://www.usgs.gov.

Table 4.5: Comparison of alternative energy technology scalable to 0.1 ZJ/year of production.
Source: C. Smith

Table 4.6: Comparison of alternative energy technology scalable to 0.1 ZJ/year of production.
Source: C. Smith

Table 4.7: Amount of Agricultural and Arable land by Geographic Region.
Source: UN Food and Agricultural Organization. http://www.fao.org/faostat.

Table 4.8: Estimated 2050 arable land needs.
Source: C. Smith

Table 4.9: Estimated 2050 pastureland land needs.
Source: C Smith

Table 4.10: Available, renewable freshwater per capita in 2017.
Source: The World Bank. https://data.worldbank.org.

Table 4.11: 2050 renewable freshwater per capita based on 2014 supply and projected 2050 population.
Source: C. Smith

Table 5.1: Impact of Population Growth, Climate Change, and Natural Resource Depletion.
Source: C. Smith

Table 5.2: Regions evaluated.
Source: C. Smith

Table 5.3: South Asia Risk Register
Source: C. Smith

Table 5.4: East Asia Risk Register
Source: C. Smith

Table 5.5: Southeast Asia Risk Register
Source: C. Smith

Table 5.6: North America Risk Register
Source: C. Smith

Table 5.7: Latin America Risk Register
Source: C. Smith

Table 5.8: European Risk Register
Source: C. Smith

Table 5.9: Russia Risk Register
Source: C. Smith

Table 5.10: Middle East Risk Register
Source: C. Smith

Table 5.11: African Risk Register
Source: C. Smith

Table 5.12: Australian Risk Register
Source: C. Smith

Table 5.13: Regional population stress.
Source: C. Smith

Table 5.14: The ten countries with the largest increase in population from 2000 to 2100.
Source: C. Smith

Table 5.15: Regional Climate Stress.
Source: C. Smith

Table 5.16: Regional Natural Resource Stress.
Source: C. Smith

Table 5.17: Governance effectiveness and stress tolerance.
Source: C. Smith

Table 5.18: Global Total Risk
Source: C. Smith

Table 6.1: Population growth rate (PGR) and total fertility rate (TFR) for South Korea, Bangladesh, Botswana, and Pakistan in 1960, 1990, and 2017.
Source: World Bank. https://data.worldbank.org.

Table 6.2: Population targets for 2050 and 2100.
Source: C. Smith

Table 6.3: African countries with high population growth and total fertility rates.
Source: World Bank. https://data.worldbank.org.

Table 6.4: Comparison of TFR in 1984 and 2017 in East Africa and West/Central Africa.
Source: World Bank. https://data.worldbank.org.

Table 6.5: Middle Eastern Countries with high population growth and total fertility rates.
Source: World Bank. https://data.worldbank.org.

Table 6.6: Middle Eastern Countries with population growth rates less than 2.0%.
Source: World Bank. https://data.worldbank.org.

Table 6.7: Comparison of NDC targets for China, India, and the United States.
Source: C. Smith

Table 6.8: Carbon dioxide emissions in Gigatonnes from the US, China, and India in 2017 compared to 2005.
Source: British Petroleum Company, *BP Statistical Review of World Energy*, 69th edition (London: British Petroleum Co., 2020).

Table 6.9: Carbon dioxide emissions in kg/$1000 GDP in the US, China, and India in 2019 compared to 2009.
Source: British Petroleum Company, *BP Statistical Review of World Energy*, 69th edition (London: British Petroleum Co., 2020).

Table 6.10: Comparison of per capita carbon dioxide emissions in the US, China, and India in 2009 and 2019.

Source: British Petroleum Company, *BP Statistical Review of World Energy*, 69th edition (London: British Petroleum Co., 2020).

Table 6.11: Comparison of Non-fossil fuel energy consumption in 2009 and 2019 for the US, China, and India.
Source: British Petroleum Company, *BP Statistical Review of World Energy*, 69th edition (London: British Petroleum Co., 2020).

Table 6.12: Comparison of Energy efficiency in 2009 and 2019 for the US, China, and India.
Source: British Petroleum Company, *BP Statistical Review of World Energy*, 69th edition (London: British Petroleum Co., 2020).

Table 6.13: Partial list of Strategies, Goals, and Objectives to meet NDC targets.
Source: C. Smith

Table 6.14: Assumptions.
Source: C. Smith

Table 6.15: Sample path to meet energy demand in 2050 and lower carbon emissions.
Source: C. Smith

Table 6.16: Calculation of carbon emissions from sample path.
Source: C. Smith

Table 6.17: Change in estimated carbon balance in 2018 and 2050.
Source: C. Smith

Table 6.18: New technologies for reducing carbon emissions.
Source: C. Smith

ABOUT THE AUTHOR

Craig R. Smith, M.D. is a physician and entrepreneur whose career has spanned academic medicine, the bio-pharmaceutical industry, and environmental biotechnology. Dr. Smith began his professional career serving on the faculty of the Department of Medicine at the Johns Hopkins University School of Medicine. Dr. Smith practiced and taught Internal Medicine and did research on the treatment of infectious diseases, the mathematical modeling of clinical outcomes, and the emerging fields of pharmocoepidemiology and pharmacoeconomics. After 16 years at Johns Hopkins, Dr. Smith joined Centocor Inc. as Vice President of Clinical Research. After directing the clinical development of the Company's therapeutic monoclonal antibody products and later serving a leadership role in Business Development, Dr. Smith returned to Baltimore and cofounded Guilford Pharmaceuticals Inc. Under his leadership Guilford grew into a fully integrated, publicly traded pharmaceutical company. After 12 years leading Guilford, Dr. Smith retired to Florida where he learned of a new method for converting carbon dioxide into ethanol using hybrid algae. Dr. Smith recognized the potential value of this technology for carbon capture and utilization and cofounded Algenol Biofuels Inc. Dr. Smith led the development of Algenol's research programs and served as Chief Operating Officer for 8 years before retiring again to travel with his wife and spend more time with his family. Dr. Smith has served on the Boards of Directors of several private and public companies, been elected to membership or Fellowship in many professional and academic societies, received numerous awards for his academic and public service, and served on many non-profit Boards of Directors.

Index

F

O

Printed in the USA
CPSIA information can be obtained
at www.ICGtesting.com
LVHW061658290823
756522LV00002B/9